ELECTROMAGNETIC FIELDS AND LIFE

ELECTROMAGNETIC FIELDS AND LIFE

A. S. PRESMAN
Department of Biophysics
Moscow University, Moscow, USSR

Translated from Russian by
F. L. SINCLAIR

Edited by
FRANK A. BROWN, JR.
Morrison Professor of Biology
Northwestern University

ℙ PLENUM PRESS • NEW YORK – LONDON • 1970

Aleksandr Samuilovich Presman completed the course of studies in the Faculty of Physics, Moscow University, in 1941. Following World War II, Presman engaged in the study of high-frequency physics. For the last two decades at the Institute of Labor Hygiene of the Academy of Medical Sciences of the USSR, the Central Institute of Health Resort Studies and Physical Therapy, and elsewhere he has studied the effects of electromagnetic fields on living organisms. Since 1966 he has given a special course on this subject in the Department of Biophysics, Faculty of Biology, Moscow University.

First Printing — February 1970
Second Printing — March 1974
Third Printing — February 1977

Library of Congress Catalog Card Number 69-12538

SBN 306-30395-7

The original Russian text, first published by Nauka Press in Moscow in 1968, has been corrected by the author for this edition. The present translation is published under an agreement with Mezhdunarodnaya Kniga, the Soviet book export agency.

Александр Самуилович Пресман

Электромагнитные поля и живая природа

© 1970 Plenum Press, New York
A Division of Plenum Publishing Corporation
227 West 17th Street, New York, N.Y. 10011

United Kingdom edition published by Plenum Press, London
A Division of Plenum Publishing Company, Ltd.
Donington House, 30 Norfolk Street, London W.C.2, England

Printed in the United States of America

PREFACE

A broad region of the electromagnetic spectrum long assumed to
have no influence on living systems under natural conditions has
been critically re-examined over the past decade. This spectral
region extends from the superhigh radio frequencies, through decreasing frequencies, to and including essentially static electric
and magnetic fields. The author of this monograph, A. S. Presman,
has reviewed not only the extensive Russian literature, but also almost equally comprehensively the non-Russian literature, dealing
with biological influences of these fields. Treated also is literature
shedding some light on possible theoretical foundations for these
phenomena. A substantial, rapidly increasing number of studies in
many laboratories and countries has now clearly established biological influences which are independent of the theoretically pre-
dictable, simple thermal effects. Indeed many of the effects are
produced by field strengths very close to those within the natural
environment.

The author has, even more importantly, set forth a novel,
imaginative general hypothesis in which it is postulated that such
electromagnetic fields normally serve as conveyors of information
from the environment to the organism, within the organism, and
among organisms. He postulates that in the course of evolution organisms have come to employ these fields in conjunction with the
well-known sensory, nervous, and endocrine systems in effecting
coordination and integration. A good case is made for the thesis
that the capacities occur in their fullest state of development only
in the organism as a whole, and are either not present, or not present in comparable form, at the molecular level.

The present convincing evidence that living systems are
steadily buffeted and stressed by the noisy fluctuations in the natural electromagnetic fields of their environment, together with the
simple experimental demonstrations that organisms behave

as specialized and very highly sensitive receptive systems for diverse parameters — strengths, frequencies, vector directions — of fields of the order of strength of the ambient natural ones, assures the subject of a permanent place among fundamental biological problems. Presman has very lucidly dealt with many general questions and issues arising from this newly acquired knowledge.

While imaginatively contemplating possible implications of this extraordinary biological responsiveness, he has maintained a very cautious, critical, and objective attitude.

This pioneering volume presents a general concept which, if future experiments continue to support it, could be one of the most fundamental and significant contributions of the 20th century. Seldom in recent times does one encounter a new scientific area for exploration with such broad implications and potential applications, so suddenly thrown into the arena for scientific scrutiny, researches, and criticism. Presman's case for his hypothesis is very ably and persuasively set forth. The postulation that normal informational roles are played by these fields for living systems at three levels is systematically developed as logical, plausible consequences and significances of the reported responsiveness of living systems to diverse experimentally applied fields, and characteristics of these responses. The resulting hypothesis is so encompassing that an impact will be felt wherever biological regulation is concerned, and this means everywhere in biology from molecular, through developmental, environmental, and behavioral, and even into the social sciences. All who read this book will thereafter view the world with an increased consciousness of possible interrelations and interactions through means hitherto considered impossible.

The author's modest hope, that the ideas which he presents will stimulate further critical discussions and researches, will undoubtedly not be a vain one. The obvious exciting possibilities for further discoveries in this relatively unmined area, the many current mysteries of biology which could find their resolution in these terms, and the unquestionably almost limitless practical applications of such additional knowledge concerning biological informational and control systems, will assure that the postulations ad-

vanced by Presman will be thoroughly explored during the coming
years.

Evanston, Illinois
January, 1969   Frank A. Brown, Jr.

FOREWORD TO AMERICAN EDITION

I am very happy to accept the proposal of the Plenum Publishing Corporation to make my book known to American scientists, who have made such a great contribution to the study of the effect of electromagnetic fields on living organisms.

In the short time which has elapsed since the publication of the original edition the results of a large number of new investigations have been published. Most of these results appear to fit in with my theory of the informational nature of the biological effects of electromagnetic fields. This is how it appears to me, but I may be guilty of unconscious bias in finding support for my view in the experimental data. Hence, I have resisted the temptation to make additions to the American edition of the book and have decided first to hear the criticism of readers in regard to the validity of the theory itself and the objectivity of its experimental substantiation.

However, while my views may (and must) be the subject of debate, I do not think that anyone will question the need for a serious investigation of the role of electromagnetic fields in the evolution and vital activity of organisms. The stimulation of the interest of scientists in this problem was the main aim of my book.

July 4, 1968   A. S. Presman

FOREWORD

Most of the external physical factors which have been implicated
in the evolution of life are of an electromagnetic nature. It has now
been established that throughout the reviewable geological period
the biosphere has been a region of electromagnetic fields and radiations of all the frequencies known to us — from slow periodic variations of the earth's magnetic and electric fields to gamma rays.

It is fundamentally possible on the basis of general considerations that any of the ranges of the electromagnetic spectrum could
have played some role in the evolution of life and are involved in
the vital processes of organisms. This has already been demonstrated for a considerable region of the spectrum — for electromagnetic radiations in the infrared to ultraviolet range (photobiology)
and from x rays to gamma rays (radiobiology).

The situation is different with the vast remaining region of
the spectrum, which includes electromagnetic fields (EmFs) of superhigh, ultrahigh, high, low, and infralow frequencies. Experimental investigations and theoretical considerations suggest that
EmFs can have a significant biological action only when their intensity is fairly high and that such action can be due to only one
process — conversion of the electromagnetic energy to heat.

Yet there is an increasing amount of reliable experimental
data which indicate that EmFs can have nonthermal effects and
that living organisms of diverse species — from unicellular organisms to man — are extremely sensitive to EmFs. Some of the discovered features of the biological action of EmFs clearly do not fit
the Procrustean bed of the heat theory. Finally, it has been found
that very weak natural EmFs can affect organisms of various species. All this indicates the necessity for a fundamentally new approach to the problem of the biological action of EmFs and for the
need to reconsider the question of the possible role of EmFs in the
vital activity of organisms.

A. S. Presman's book is the first attempt at such an approach to the problem on the basis of the concept of the informational role of EmFs in the evolution and vital activity of organisms. It should be pointed out straight away that, on the whole, the author has coped very successfully with this problem.

The introduction postulates the existence of three kinds of "biological activity of EmFs" — the effect of natural environment EmFs on the regulation of vital processes, the role of internal fields in the organism in the coordination of physiological processes, and the interaction between organisms by means of EmFs. The author clearly demonstrates that it is not sufficient to consider only the energetic aspect of the interaction of EmFs with biological systems and indicates ways of investigating the informational functions of these fields in living nature.

In the light of these basic viewpoints the book gives a thorough account of the physical aspects of the interaction of EmFs with biological systems and critically analyzes practically all the known information on the biological action of artificially produced and natural EmFs. It should be noted that Presman has been able to approach this broad review from various standpoints of natural sciences: He considers the biological effects of EmFs from the physical, biological, and cybernetic viewpoints. Hence, the generalizations and conclusions regarding the main features of the biological action of EmFs are impartial and quite objective.

The author puts forward some interesting and quite feasible hypotheses on the significant role of EmFs in the still unexplained mechanisms of some interconnections within organisms, between organisms, and between organisms and the environment.

Of course, not all the views expressed in the book are sufficiently well substantiated and some are quite controversial. This is quite understandable, however, in the first formulation of a new biological problem.

The author's concept of the informational functions of EmFs in living organisms and the hypotheses expressed in this respect will undoubtedly rouse interest in a wide circle of readers and stimulate discussion on this problem. In addition, and this is very important, the book will certainly be very useful as the first extensive review and analysis of the experimental and theoretical investigations of the biological action of EmFs, particularly since the author considers not only the heuristic aspect of the problem, but also its various practical applications. Finally, one of the merits of the book is the lucidity of the exposition for specialists in the most diverse fields — biologists and biophysicists, doctors and physiologists, physicists and radio engineers.

Academician V. V. Parin

CONTENTS

Part Two

EXPERIMENTAL INVESTIGATIONS OF
THE BIOLOGICAL ACTION OF
ELECTROMAGNETIC FIELDS

Part Three

ROLE OF ELECTROMAGNETIC
FIELDS IN THE REGULATION OF THE
VITAL ACTIVITY OF ORGANISMS

INTRODUCTION

Since the above lines were written science has learned much more
about the electromagnetic radiations which have been present in
the biosphere — the region which living organisms inhabit — throughout the whole epoch of evolution. Scientists have discovered more
and more new natural electromagnetic radiations in various ranges
of the electromagnetic spectrum. To the long-investigated range
of solar radiations — infrared to ultraviolet — we have now added
ionizing radiations (x rays and γ rays) of cosmic and terrestrial
origin. In the remaining lower-frequency region of the electromagnetic spectrum, the discovery of the slow periodic variations
(seasonal, monthly, diurnal) of the earth's magnetic and electric
fields was followed by the discovery of the short-period fluctuations
of the earth's magnetic field with frequencies extending to hundreds
of hertz. The investigation of atmospheric discharges has shown that
the electromagnetic radiations produced in this case cover a wide
range of wavelengths — from superlong to ultrashort. Finally, radio
emission in the range of meter to millimeter waves from the sun and
the galaxies has been discovered.

Thus, it has now been established that periodic electromagnetic processes with frequencies distributed throughout the known

1

electromagnetic spectrum are a permanent feature of the biosphere. Hence, it can be postulated *a priori* that any region of this natural electromagnetic spectrum might have played some role in the evolution of living organisms and that this will be reflected in some way or other in their vital processes. We can say that any region of the spectrum is biologically active in some degree. In the general case three kinds of activity are probable:

1. The effect of electromagnetic processes taking place in the environment on the functioning of living organisms.
2. The role of electromagnetic processes taking place within organisms in the vital activity of organisms.
3. Electromagnetic interconnections between organisms.

For the spectral region where $h\nu > kT$ (at temperatures characteristic of living organisms), i.e., from the infrared range to gamma rays, these kinds of biological activity have already been discovered to some extent. It is known that photobiology is concerned not only with the reaction of living organisms to infrared, visible light, and ultraviolet rays, but also with the role of radiations of these ranges in processes occurring with organisms and "light-mediated interaction" between organisms. Radiobiology is concerned at present only with the biological action of x rays and gamma rays,* but "internal radiation" has also been discovered, although its role in biological processes has not yet been elucidated. Some recent investigations indicate the possibility of "radiation-mediated interactions" between organisms.

The situation is different with the remaining vast region of the electromagnetic spectrum, where $h\nu < kT$; this region includes the ranges from super-high to ultralow frequencies, right down to "zero frequency" (constant electric and magnetic fields). For convenience of exposition we will henceforth call this whole region of the spectrum "electromagnetic fields"[†] or EmFs. The problem of the biological activity of EmFs as a whole has only begun to take shape in recent years, although investigations of individual aspects of this problem has been carried out some time ago.

---

* We exclude here the corpuscular kinds of radiation (α- and β-rays, neutron fluxes), since we are dealing only with electromagnetic radiations.

† This term is arbitrary, of course, since all the rest of the spectrum also consists of electromagnetic fields.

Galvani's experiments in the 18th century initiated the development of electrophysiology (electrobiology), which deals with the reactions of living organisms to electric stimuli and with electric phenomena in organisms themselves. Using the terminology which we have adopted we can say that electrophysiology is the study of the first and second kinds of biological activity of EmFs, mainly in the low-frequency range and mainly in the case of contact application (or withdrawal) of electric energy. Galvani, however, was also first to discover a long-distance electric effect when he observed the contraction of the muscle in a frog nerve—muscle preparation situated at some distance from the spark of an electrostatic machine. Such experiments on electric stimulation of nerve at a distance were not taken up again until the end of the 19th century. At this time there also appeared the first reports of the biological effects of a magnetic field and high-frequency fields. In 1900-1901 there appeared V. Ya. Danilevskii's two-volume monograph dealing with the main experimental and theoretical principles of the biological action of EmFs on biological objects — from cells to entire organisms.

In the thirties and forties research on the effect of high- and ultrahigh-frequency EmFs on the human and animal organism was greatly extended (Libezin, 1936). Frenkel and Tatarinov, 1939; Proceedings of the Conference on the Application of UHF in Medicine, 1940; Shcherbak, 1936). The researchers encountered two kinds of biological effects from EmFs. There was a clear demonstration of the thermal effect — heating of the tissues of the organism or the biological substance in vitro due to EmFs of fairly high intensities; there were less definite effects which could not be attributed entirely to heating and also effects due to low-intensity EmFs, where no heating of the tissues was detected. Such "nonthermal" effects (which were often unwisely called "specific" effects) were observed mainly in the reactions of entire organisms, more rarely in isolated organs, and hardly ever in experiments with macromolecular solutions in vitro.

A possible role of natural EmFs in living nature has been categorically rejected on the basis of the following theoretical considerations.

Any possible effect of EmFs on biological objects, as on nonliving objects similar to them in electrophysical properties, must be due to particular energetic interactions of the EmF with the substance, i.e., conversion of the electromagnetic energy to other forms such that the resultant effect depends on the effective energy of the EmF. But the conversion of the EmF energy into other forms of energy in living tissues can involve only the same processes as in electrolyte solutions of corresponding composition. As regards the effect of a constant magnetic field, living tissues must be regarded as a weakly magnetic substance. Moreover, the energies of quanta in the EmF region are sufficient only to cause vibrations of charged particles as a whole — ions, dipole molecules, and colloidal micelles — in biological media. Such processes represent the only possible interaction of EmFs with biological media and result in the conversion of the electromagnetic energy to heat energy. However, for the induction of any biologically significant thermal effect the strengths of the acting EmFs must be several orders greater than those of natural environment EmFs of corresponding frequency. The biological effects of a constant magnetic field can be due to the orientation of paramagnetic and diamagnetic molecules. Such effects are possible only if the energy of the magnetic field, calculated per molecule, exceeds kT. For this the intensity of the field must be at least ten thousand times greater than that of the geomagnetic field.

The experimental data indicating changes in physiological processes in entire organisms due to exposure to much weaker EmFs were regarded as unconvincing for the following reasons: Firstly, organisms can respond with similar changes to diverse external factors and, hence, there are no grounds for postulating a "specific" biological effect of EmFs; secondly, if there was such an effect it would certainly be detected in experiments with physicochemically homogeneous media or, at least, in experiments with very simple biological systems, but such attempts were unsuccessful. Thus, physicists concluded that weak EmFs were incapable of producing biological effects.

In spite of these categorical conclusions biologists continued with experimental attempts to detect biological effects due to EmFs and constant magnetic fields with strengths much lower than the theoretically predicted effective values. Within the last ten years these attempts have produced successful results, which give grounds for believing that natural EmFs have probably been implicated in

the evolution of life and play a significant role in the vital activity
of organisms. One cannot help recalling in this connection the
words of Szent-Gyorgyi (1960) that "the biologist depends on the
judgement of the physicist, but must be rather cautious when told
that this or that is improbable."

Biological investigations have shown that organisms of the
most diverse kinds — from unicellular organisms to man — are
sensitive to a constant magnetic field and EmFs of different frequencies with an effective energy tens of orders (!) less than the
theoretically estimated effective level. Hence, such estimates,
based on the concept of the energetic interaction of EmFs with the
biological substance, obviously lack a sound basis. This concept
is also contradicted by the fact that instead of the predicted proportional relationship between the biological effects and the intensity of the acting EmFs quite different relationships have been experimentally established. It has been found that in some cases the
reactions of living organisms to EmFs occur only at certain "optimum" intensities, in other cases the effects increase when the intensity of acting EmF is reduced, and in other cases the reactions
to low and high intensities are of opposite nature. "Cumulative"
biological effects produced by repeated exposure to EmFs well below the effective threshold for a single exposure have also been
observed. Finally, the concept of energetic interaction is contradicted by the fact that for the same average EmF energy absorbed
in the tissues of organisms the nature of the reactions depends
considerably on the modulation of the EmF, on the directions of the
electric and magnetic vectors of the EmF relative to the animal's
body axis, on the localization of the exposure, and so on.

This has given rise to the need for a fundamentally new
theoretical approach to the problem of the biological activity of
EmFs — a theory which will not only be consistent with the experimental data but will provide a basis for their interpretation and an
elucidation of the particular mechanisms involved. Such an approach can be based on information theory.

The application of this theory to biology has shown that, in
addition to energetic interactions, informational interactions play a significant (if not the main) role in biological processes. Such interactions entail the conversion of information, its

transmission, coding, and storage. The biological effects due to these interactions do not depend on the amount of energy introduced into the particular system, but on the amount of information introduced into it. The information-carrying signal merely causes the redistribution of the energy in the system itself and regulates the processes occurring in it. If the sensitivity of the receiving systems is sufficiently high very little energy is required for information transfer. The information can be built up in the system by the repetition of weak signals.

The informational aspect of the interaction of EmFs with biological objects must be taken into account even in the consideration of such obviously energetic effects as thermal effects. These effects are due not only to an increase in the kinetic energy of c h a o t i c motion of the molecules, as in normal heating (convection, infrared radiation), but also to c o h e r e n t vibrations of the molecules and ions with the same frequency as the acting EmF. In biological systems with oriented molecular layers and ordered motions of the microparticles such an imposed rhythm can be regarded in general as the introduction of "harmful" information into the system.

The information approach to the biological action of EmFs is even more essential in the case of very low intensities, where any energetic effects are theoretically impossible. Informational interactions of EmFs with biological systems can account for the high sensitivity of living organisms to EmFs, the specific relationship between the biological effects of EmFs and their intensity and modulation, and the cumulative effect of EmFs.

It is a valid assumption that all these special features of the reactions of living organisms to EmFs are associated with certain biological systems formed in the process of evolution for the reception of information from the environment. This hypothesis has already received experimental verification. It has been found that the periodic variations of natural environmental EmFs have a regulating effect on vital functions — on the rhythms of the main physiological processes, on the ability of animals to orient themselves in space, on multiplication in populations, and so on. In the living organism the systems for receiving information transmitted by EmFs are reliably shielded from natural electromagnetic interference, but in pathological states spontaneous variations of EmFs (solar flares, lightning discharges) upset the regulation of physiological processes.

It appears that in the process of evolution living nature has used EmFs to obtain information about the changes in the environment. The EmF, of course, is the most reliable information-carrier among geophysical factors. By means of EmFs information can be transmitted (in the appropriate frequency ranges) through any medium inhabited by living organisms and in any meteorological conditions — during the polar day or night, in river and sea water, within the earth's crust and, finally, in the tissues of organisms themselves.

Thus, the first kind of biological activity of EmFs — the effect of environmental EmFs on the functioning of living organisms — acquires reality in the light of the results of biological investigations and their interpretation on the basis of the concept of the informational interactions of EmFs with biological systems formed in the process of evolution.

Experimental data which indicate the existence of the second kind of biological activity of EmFs — their implication in informational interconnections within the organism — have now been obtained. There is not only the already known transmission of information along the nerves by bioelectric pulses with a frequency spectrum extending to thousands of hertz, but also the long-range interactions effected by EmFs of very different frequencies — from infrared to superhigh frequency (we will henceforth call these EmF-interactions or EmF-connections). They are manifested in the electromagnetic interrelationships between macromolecules, in the synchronization of the electromagnetic vibrations in assemblies of macromolecules and groups of cells, and so on. It is hoped that the concept of EmF-connections may clarify the hitherto baffling nature of some long-established selective interactions between cells and macromolecules.

Finally, there are data which suggest the existence of the third kind of biological activity of EmFs — informational interconnections between organisms. On one hand, animals are particularly sensitive to EmFs with parameters (frequency, intensity, pulse modulation) in a fairly narrow range. On the other hand, it has been found that living organisms are sources of EmFs of various

ranges — from infralow to superhigh frequency. Finally, exposure
of human beings to EmFs gives rise to definite sensations — visual,
auditory, and tactile. The concept of informational EmF-connections
in the animal world may prove fruitful in several cases where
there is clearly an exchange of signals between animals, but the
physical nature of the signaling process has not yet been determined.

An analysis of the experimental data which indicate the appearance of these kinds of biological activity of EmFs leads to the
conclusion that it is useless to attempt to seek the causes of such
properties of life o n l y at the molecular level or to proceed from
a consideration of the processes arising at this level due to EmFs.
This is most obvious from investigations of the effect of external
EmFs on biological objects.

It has been found, for instance, that entire organisms are
most sensitive to EmFs, isolated organs and cells are less sensitive, and solutions of macromolecules are even less sensitive.
Significant differences are observed in the reaction of one biological system (molecular, cellular, organic, or systemic) to EmFs in
relation to the conditions in which the action takes place — whether
it is in the entire organism or in the isolated state. In these two
cases there is also a difference in the nature of the relationship
between the reaction of the system and the parameters of the EmF.

All this indicates that the systems which are particularly
sensitive to EmFs have presumably been formed in the process of
evolution only at the macroscopic level, beginning, at the lowest,
from ordered macromolecular assemblies. In other words, the
property of receiving weak natural EmFs appears only at the level
of fairly complexly organized biological systems and this property
is probably manifested to the full extent only in the entire organism.

The appearance of enhanced sensitivity to EmFs only in fairly
complexly organized biological systems, can be regarded as one of
the manifestations of the specific nature of life — its "organization." We cite some current views on this subject.

Shmal'gauzen (1964) says ". . . the functions of a higher system are not a summation of the activity of lower systems, but their
i n t e g r a t i o n. Each higher system reveals its own q u a l i t a t i v e

specificity, which is created only by the organization of this higher system."

Szent-Gyorgyi (1960) says "One of the basic principles of
life is "organization" by which we mean that if two things are put
together something new is born, the qualities of which are not addi-
tive and cannot be expressed in terms of the qualities of the consti-
tuents."

Schmitt (1959) speaks of "hierarchies of organization and with
the properties that are characteristic of systems . . . at each particular level of organizational complexity, viz., molecular, macromolecular, subcellular, cellular, supercellular, organismic, and
superorganismic." Further,". . . there is a formidable "black box"
between the molecular effectors, as studied in the model system,
and the final behavior of the cells or organism under study."

It has been pointed out on occasion that such integration
occurs also in nonliving nature — in the systems of elementary
particles forming atoms and in the combination of atoms to form
molecules. At these levels, however, there is no difference between nonliving and living matter: The atoms and molecules are
organized in the same way in each case. The difference appears
only at the "supermolecular" level — in the specific nature of the
organization of the macromolecules and their assemblies, and in
the hierarchy of the systems comprising the living organism. Here
we encounter the integration of properties which has no analogy in
nonliving nature.

We have made this excursion into the problem of the specificity of life, since up till now the common approach to the investigation of the biological effect of EmFs has been one in which effects at the molecular level in physicochemically homogeneous systems are first investigated and only after this are the more complexly organized structures examined. There is still a sceptical
(and usually negative) attitude to any results of experiments on entire organisms and even on isolated organs. Such views are the
result of excessive enthusiasm over the undoubtedly significant
successes of "molecular biophysics" and the neglect of the histori-
cal fact that the investigation of the interaction of living organisms
with any external factor began with observation of the reactions of

the entire organism and only after this were the systems involved
in these reactions revealed.

Yet the reverse procedure is the practical one: To investigate first the reactions of the entire organism to EmFs, and then
to go on from there to determine the simplest level of organization
at which it is still possible to detect the changes which are ultimately responsible for the particular reaction of the organism.
The prime cause of the reaction need not necessarily be due to processes at the molecular level — it may be a feature of some macroscopic level of organization.   There is nothing paradoxical in
this — we encounter such a situation in the examination of "organized" machine systems, which are also characterized by the presence of a hierarchy of order and the appearance of new properties
with increase in complexity.*

This can be illustrated even by the example of such a relatively simple system as the oscillatory circuit. The specific feature of this circuit is its particularly high sensitivity to an EmF of
a particular (resonant) frequency. Of course, the electromagnetic
oscillations induced in the circuit are due to microprocesses —
motion of electrons in the wires and the polarization of the molecules in the dielectric of the capacitor. But attempts to discover
the mechanism of resonance at this level will be in vain — it is not
there. Nor can we detect this property by considering the processes
in separate macroscopic elements of the circuit — the capacitor
and the induction coil. The property of resonance is a feature only
of the whole organized system — the circuit as a whole.

Thus, the phenomenological approach to the investigation of
the systems and processes involved in the reactions of biological
objects to EmFs is the practical one at present. We can determine only the regular relationships between the "input" and "output" characteristics of the system, i.e., between the operating factors and the reactions. It is quite in order to simulate such systems and processes by means of equivalent electromagnetic "organized systems", exhibiting corresponding input and output charac-

*This may be the result of a historically manifested law which McCulloch (1962) ex-
pounds in the following way: "But aside from sources of power, and from the wheel,
most of what we have done has been an imitation (of animals)."

teristics, the relationship between which is described within the framework of the macroscopic parameters. In this way we can, for instance, approach an explanation of the high sensitivity of biological systems to EmFs, i.e., their ability to receive the information carried by EmFs of very low intensity (even below the noise level).

Information theory indicates the possiblity of "spatial summation" of electromagnetic informational signals received simultaneously by n elements and of "temporal summation" of signals repeated n times. In either case the total signal-to-noise ratio is increased by a factor $\sqrt{n}$. Hence, if n is sufficiently large the reception of information with a signal intensity below the noise level is possible. Reception has been effected in technical systems of such kind. The two types of summation are also found in living organisms — in the visual organs for electromagnetic light waves (Glezer and Tsukerman, 1961) and in nerve cells for low-frequency electromagnetic signals. Hence, the existence of such biological systems for the reception of EmFs of other frequency ranges is probable.

The main information on the input and output characteristics of biological systems sensitive to EmFs has been obtained mainly from investigations of the reactions of entire organisms to EmFs. However, to seek certain specific reactions is "to pursue a will-o-the-wisp": to adequate informational action of Em Fs and to electromagnetic noise the reactions are nonspecific — the same reactions can be produced by other external factors.

For instance, regulation of the daily rhythm of physiological processes is effected by various geophysical factors, probably equally. The role of periodic environmental EmFs in this regulation can be determined only by a correlation analysis, either by artificially maintaining other geophysical factors constant, or by investigating changes in the rhythm of physiological processes when the environmental EmFs are artificially attenuated.

The disturbances of physiological processes — changes in intensity, direction, etc. — due to electromagnetic interference must also be nonspecific. The same kind of disturbances can be produced by the action of other factors — mechanical, chemical, biological, etc. The disturbance of the function of any organized system is nonspecific in the sense that its nature is often independent of the type of interference. For instance, the same noise can be produced at the output of a radio receiver by radio interference (natural or artificial) or by faulty contacts — mechanical or chemical.

These remarks on the approach to the investigation of the biological action of EmFs apply also the EmF-interconnections within the organism. Interactions between macromolecules and between cells and, even more so, between more complex biological systems cannot be understood at present without regard to the role of "organization" at all these levels and without regard to the differences in the course of the corresponding processes in the organism and in systems isolated from it.

As regards the investigation of the problem of EmF-interconnections between animals, even investigations of entire organisms may be inadequate. Since associations in groups, communities, populations, and biocenoses exhibit their own specific features due to "coalition" (von Foerster, 1962) — "an aggregate of elements which jointly can do things which all of them separately could never achieve". Hence, the investigation of the EmF-interconnections in animal associations will require an analysis of a vast amount of material collected by zoologists, ecologists, and ethologists. In this respect the method of empirical generalization, proposed by Vernadskii (1926), may prove to be very effective. This method, which is based on inductively obtained facts, does not go beyond the limits of these facts and is not concerned with the consistency or inconsistency of the obtained conclusion with other existing ideas on nature. In this respect empirical generalization does not differ from scientifically established fact: Their agreement with our scientific ideas on nature does not interest us, their contradiction of them is a scientific discovery.

Such briefly are the general arguments and experimental evidence which have led us to pose the problem of the role of EmFs in life. In fact, in addition to previously developed fields of biology — photobiology and radiobiology, which are concerned with the biological activity of electromagnetic radiations from infrared to gamma rays, there is now developing a new branch of biology dealing with the biological activity of the remaining region of the electromagnetic spectrum — from superhigh to "zero" frequency (constant fields). We suggested earlier (Presman, 1965a) that this branch, which brings together a wide circle of problems relating to various manifestations of the biological activity of EmFs, should be called e l e c t r o m a g n e t i c   b i o l o g y (for lack of a more suitable term).

The basic tenets in the formulation of the problem are: 1) the concept of the i n f o r m a t i o n a l   f u n c t i o n s of EmFs in life at all levels of its hierarchical organization and 2) the conviction that the investigation of these functions of EmFs must proceed from c o m p l e x   b i o l o g i c a l   s y s t e m s   a n d   p r o c e s s e s   t o   m o r e and   m o r e   s i m p l e   s y s t e m s   a n d   p r o c e s s e s .

The aim of this book is to substantiate this concept and methodology from the abundant experimental material on the diverse biological effects of EmFs. I have tried to carry out this analysis as objectively as possible, but I admit that, being the author of this concept, I have found it difficult to avoid partiality in the assessment of the experimental data which I have examined and the derivation of conclusions from them.

In such a case it is particularly valuable to have the opinion of impartial scientists as to how far the author has managed to retain sufficient objectivity in the exposition and substantiation of his views. For such a critical evaluation of the manuscript of this book, and for valuable advice and comments I am deeply grateful to Academician V. V. Parin and Professors L. A. Blyumenfel'd, P. I. Gulyaev, and S. É. Shnol'.

Part One

PHYSICAL PRINCIPLES
AND EXPERIMENTAL METHODS
OF INVESTIGATION
OF THE BIOLOGICAL ACTION
OF ELECTROMAGNETIC FIELDS

Chapter 1

PHYSICAL CHARACTERISTICS OF
ELECTROMAGNETIC FIELDS

Table 1 gives the region of the electromagnetic spectrum with which this book is concerned and which we have arbitrarily called EmFs. In addition to currently adopted designations of frequency and wavelength ranges, the table also gives earlier designation (HF, UHF, microwaves), since these terms still occur frequently in accounts of research on the biological action of EmFs; the values of the ratio $h\nu/kT$ for the frequency ranges in the considered region of the spectrum are also given in the table.

This part gives an account of the characteristics of the main parameters of natural and artificial sources of EmFs of interest from the biological viewpoint and a description of methods of determining these parameters. We also discuss the electric and magnetic properties of biological media, the laws governing the absorption and conversion of EmF energy in the tissues of living organisms and, finally, the main experimental techniques in biological investigations of EmFs.

TABLE 1. Region of Electromagnetic Spectrum from Infralow to Superhigh Frequencies, in Which $h\nu < kT$

Wave range	Wavelength, cm	Frequency, Hz	$h\nu/kT$	Frequency range
Low-frequency waves	10^{13}	$3\cdot10^{-3}$	$4.5\cdot10^{-16}$	Infralow
Low-frequency waves	10^{12}	$3\cdot10^{-2}$		Low
Low-frequency waves	10^{11}	$3\cdot10^{-1}$		
Low-frequency waves	10^{10}	3		Industrial
Low-frequency waves	10^{9}	$3\cdot10^{1}$		
Low-frequency waves	10^{8}	$3\cdot10^{2}$		Audio
Low-frequency waves	10^{7}	$3\cdot10^{3}$		
Radio waves — Long	10^{6}	$3\cdot10^{4}$	$4.5\cdot10^{-9}$	High (HF)
Radio waves — Medium	10^{5}	$3\cdot10^{5}$		
Radio waves — Intermediate	10^{4}	$3\cdot10^{6}$		
Radio waves — Short	10^{3}	$3\cdot10^{7}$		
Ultraradio waves (Microwaves) — Meter	10^{2}	$3\cdot10^{8}$	$4.5\cdot10^{-5}$	Ultrahigh (UHF)
Ultraradio waves (Microwaves) — Decimeter	10	$3\cdot10^{9}$		
Ultraradio waves (Microwaves) — Centimeter	1	$3\cdot10^{10}$		Superhigh (SHF)
Ultraradio waves (Microwaves) — Millimeter	$0{,}1$	$3\cdot10^{11}$		
Ultraradio waves (Microwaves) — Transitional	$0{,}01$	$3\cdot10^{12}$	$0{,}45$	

However, before turning to the discussion of these questions we must give at least a brief account of the main features of electromagnetic phenomena, the physical quantities characterizing these phenomena, and the most important relationships between these quantities.

1.1. Electric and Magnetic Fields

The electric field created in the vicinity of an electrically charged body is a vector quantity. The absolute value of this vector — the force with which the field acts on a unit charge situated at a particular point in space — is called the electric field strength E and is measured in volts per meter (V/m). The direction of the vector is the direction in which a positive charge moves in this field. The trajectories of the motion of this charge, placed at one point or another in the field, are called the electric lines of force.

The magnetic field formed around a conductor carrying a current or in the vicinity of a permanent magnet is also a vector quantity. The magnetic field strength H is the force with which the field acts on an element of current situated at a particular point. The value of H is measured in amperes per meter (A/m), but a unit $10^3/4\pi$ times larger — the oersted (Oe) — is often used. The trajectories of the motion of an element of current (or the orientations of an elementary magnet) in a magnetic field are called the magnetic lines of force.

1.2. Electric and Magnetic Properties

of a Medium

The electric properties of a medium are characterized by two quantities:

1) the dielectric constant, expressed as a relative quantity ε' (dimensionless) or an absolute quantity ε, measured in farads per meter (F/m)

$$\varepsilon' = \frac{\varepsilon}{\varepsilon_0},\tag{1}$$

where $\varepsilon_0 = 8.854 \cdot 10^{-12}$ F/m is the absolute value of the dielectric constant in vacuo (practically the same value as in air);

2) the specific electrical conductivity σ, measured in siemens per meter (S/m) or in units hundred times larger — mho/cm. The reciprocal of this quantity — the electrical resistivity ρ — is often used.

The magnetic properties of a medium are also characterized by two quantities:

1) the magnetic permeability, expressed as a relative (dimensionless) quantity μ' or as an absolute quantity μ, measured in henries per meter (H/m)

$$\mu' = \frac{\mu}{\mu_0},\qquad(2)$$

where $\mu_0 = 1.127 \cdot 10^{-6}$ H/m is the absolute value in vacuo (which is approximately the same for air);

2) the permeance g_m, measured in webers per ampere (Wb/A), or the reluctance r_m — the reciprocal of g_m.

1.3. The Electric Field and Electromagnetic Waves

Any change in an electric field is alway accompanied by the appearance of a magnetic field and, conversely, any change in a magnetic field leads to the appearance of an electric field. Fields such as this, which are related to one another and can be converted to one another, are called electromagnetic fields, the main parameters of which are: the frequency f of the oscillations (or the period $T = 1/f$), the amplitude E (or H), and the phase φ, which defines the state of the oscillatory process at any instant. The frequency is expressed as the number of oscillations per second in hertz (Hz), kilohertz (1 kHz = 10^3 Hz), megahertz (1 MHz = 10^6 Hz), and gigahertz (1 GHz = 10^9 Hz). The quantity $\omega = 2\pi f$ (the angular frequency) is also used. The phase is expressed in degrees or relative units — multiples of π.

An electromagnetic field is propagated in the form of electromagnetic waves, the main parameters of which are the wavelength λ, the frequency f, and the velocity of propagation $v = c/\sqrt{\varepsilon'\mu'}$, which are connected by the relationship:

$$\lambda = \frac{c}{f\sqrt{\varepsilon'\mu'}},\qquad(3)$$

where $c = 3 \cdot 10^8$ m/sec is the velocity of light in vacuo (practically the same as in air, where $\mu' = \varepsilon' = 1$).

Wave formation takes place in the w a v e   z o n e at a distance of more than λ from the source. In this zone E and H vary in phase and their mean values during the cycle are connected by the relationship

$$E = \sqrt{\tfrac{\mu_0}{\varepsilon_0}}\, H = 377 \cdot H. \tag{4}$$

At smaller distances — in the i n d u c t i o n   z o n e — E and H vary out of phase and decrease rapidly with increasing distance from the source (in inverse proportion to the square and cube of the distance), and there can be any relationship between their mean values. In the induction zone the energy is converted alternately to an electric and magnetic field. Hence, E and H are evaluated separately here. In the wave zone energy is radiated and this radiation is usually evaluated in units of p o w e r   f l u x   d e n s i t y S (the Poynting vector), which is expressed in W/m^2 [units less by a factor of 10^4 (W/cm^2) or 10^7 (mW/cm^2) are often used]:

$$S = E \cdot H. \tag{5}$$

The direction of this flux is determined by the "corkscrew rule": If the handle of a corkscrew is rotated in the direction away from the vector E towards the vector H (which are perpendicular to one another in an electromagnetic field) the direction of its penetration is the direction of S.

The power flux density at a distance R from the source can be estimated from the total radiated power P:

$$S = \frac{P}{4\pi R^2}. \tag{6}$$

In the case of directed radiation this value must be multiplied by a d i r e c t i v i t y   f a c t o r, which is determined from the parameters of the radiator.

There are two types of electromagnetic oscillations which occur most frequently — h a r m o n i c, in which E and H vary according to a sine or cosine law, and m o d u l a t e d, in which the amplitude, frequency, or phase vary additionally in accordance with a particular law. We correspondingly speak of sinusoidal electromagnetic waves and modulated waves.

The interaction of two (or several) waves gives rise to in-terference — an increase or decrease in the amplitude of the resultant wave according to the relationship of the phases in the wave space.

Of particular interest for the subsequent account is pulse modulation, where electromagnetic pulses of short duration τ are separated by long pauses. The frequency band corresponding to the pulse has a value of the order $1/\tau$, and the connection between the power in the pulse and the mean power is given by the relation-ship

$$P_p = \frac{P_m}{F \cdot \tau},$$ (7)

where F is the pulse repetition rate (expressed in Hz or pulses/sec). The quantity $1/F\tau$ is called the off-duty factor.

If the phase difference between interfering waves is constant (at least during the time of observation) they are said to be cohe-rent. In this case the resultant intensity is more or less than the sum of the initial intensities, depending on the phase difference. If the phase difference varies in a random manner the waves are incoherent and the resultant mean intensity is equal to the sum of the initial intensities.

We will mention one other characteristic of waves — polar-ization, which includes cases where the directions of the vectors E and H in space are constant (linear polarization) or vary according to a particular law (elliptic and circular polar-ization). The plane passing through the direction of propagation of the wave and the direction of polarization is called the plane of polarization.

1.4. Interaction of EmFs with a

Physical Medium

We will be interested in those physical media which are sim-ilar in their electric and magnetic properties to the the tissues of living organisms. Such media are electrolyte solutions containing protein molecules with weakly diamagnetic and paramagnetic pro-perties and an electric polarity characterized by the dipole moment.

The action of an electrostatic field in such media causes the movement of "free" electric charges (electrons, ions, and other charged particles), polarization, i.e., the displacement of "bound" charges (electrons in atoms, atoms in molecules), and the orientation of molecules with a constant dipole moment (protein and water molecules). A magnetostatic field causes orientation of diamagnetic and paramagnetic molecules. Moving electric charges are acted on by such a field with a force given by the equation

$$F = qvH, \tag{8}$$

where q is the electric charge, v is its velocity, and H is the magnetic field strength.

The direction of the force F is determined by the corkscrew rule: When the handle of the corkscrew is rotated away from the direction of v towards the direction of H the corkscrew moves in the direction of F.

The action of alternating EmFs in the considered medium will give rise to processes of two main types — oscillations of free charges and rotations of dipole molecules in accordance with the frequency of the EmFs. Since the medium has electric resistance and viscosity these two processes entail the loss of EmF energy: In the first case it is called c o n d u c t i o n   l o s s and in the second case d i e l e c t r i c   l o s s.

The loss of either kind and its share in the total absorption of EmF energy in the medium depends, firstly, on its electric parameters — the electrical conductivity and the dielectric constant — and, secondly, on the frequency of the acting EmFs.

The relationship between the conduction loss and the dielectric loss is usually expressed by the loss tangent tan δ or the complex dielectric constant ε^*. These quantities are connected with one another by the following relationships:

$$\tan \delta = \frac{\varepsilon''}{\varepsilon'} = \frac{\sigma}{\omega \varepsilon' \varepsilon_0}, \tag{9a}$$

$$\varepsilon^* = (\varepsilon' - j\varepsilon') \varepsilon_0, \tag{9b}$$

where ε'' is the l o s s   f a c t o r and σ is the conductance, which takes into account both kinds of loss.

The medium is regarded as conducting if the conduction loss in it greatly exceeds the dielectric loss, i.e., when tan $\delta \gg 1$; as semiconducting when the two kinds of loss are approximately the same, i.e., when tan $\delta \simeq 1$; and as dielectric, if the dielectric loss greatly exceeds the conduction loss, i.e., tan $\delta \ll 1$.

As equation (9a) shows, the value of tan δ depends on the frequency. Hence, one medium can behave as conducting with respect to an EmF of one frequency range, exhibit semiconducting properties in an EmF of another range and, finally, manifest dielectric properties in relation to an EmF of a third frequency range. For instance, sea water (which is similar in its mineral composition to physiological saline) behaves as a conductor towards EmFs with frequencies below MHz (tan δ = 100), as a dielectric at frequencies above 10 GHz (tan δ = 0.1), and as a semiconductor in the frequency region close to 1 GHz (tan δ = 1).

The EmF power dissipated in a conducting medium per unit volume is independent of the frequency and is given by the relationship

$$P_c = \sigma E^2. \tag{10}$$

The power dissipated per unit volume of a dielectric medium depends on the frequency, as can be seen from the expression

$$P_d = \omega \varepsilon' \varepsilon_0 \tan \delta E^2. \tag{11}$$

In addition, the quantity ε^* itself varies with the frequency (dispersion), since any polarization involves transient r e l a x a-t i o n   p r o c e s s e s. This means that charging and discharging processes are not instantaneous, but take a certain finite time — the r e l a x a t i o n   t i m e τ, which depends on the structure of the polarizable elements, the viscosity of the medium, and its temperature. The frequency dependence of ε^* due to this is expressed in the following way:

$$\varepsilon' = \varepsilon'_\infty + \frac{\varepsilon'_s - \varepsilon'_\infty}{1 + (\omega\tau)^2}, \quad \varepsilon'' = \frac{(\varepsilon'_s - \varepsilon'_\infty)\,\omega\tau}{1 + (\omega\tau)^2}$$
$$\sigma = \sigma_s + \frac{(\sigma_\infty - \sigma_s)\,(\omega\tau)^2}{1 + (\omega\tau)^2}, \tag{12}$$

where the subscript s indicates the values at very low frequencies and the subscript ∞ indicates the values at very high frequencies.

These equations apply to three types of relaxation processes. The first type is the relaxation of molecules which have a constant dipole moment, when equations (12) are called Debye equations and τ is determined by the viscosity of the medium η, the radius of the molecule a, and the absolute temperature T:

$$\tau = \frac{4\pi a^3 \eta}{kT},\qquad (13)$$

where k is the Boltzmann constant.

The second type of relaxation relates to an inhomogeneous structure — a suspension of spherical particles with a dielectric constant ε_i' and conductivity σ_i, occupying a fraction p of the volume of the solution (ε_a' and σ_a). In this case equations (12) are called Maxwell-Wagner equations and have the following parameters:

$$\tau = \varepsilon_0 \frac{\varepsilon_i' + 2\varepsilon_a'}{\sigma_i + 2\sigma_a};\quad \varepsilon_s - \varepsilon_\infty = 9_p \frac{(\varepsilon_i' \sigma_a - \varepsilon_a' \sigma_i)^2}{(\varepsilon_i' - 2\sigma_a)(\sigma_i + 2\sigma_a)^2}.\qquad (13a)$$

The third type is the relaxation associated with polarization at the interfaces when the ion-containing medium contains particles of different size with surface electric charges. For this case there are several equations of type (12) for different values of τ.

Figure 1 shows graphs of the variation of ε', tan δ, and σ with frequency for the relaxation mechanisms represented by the Debye equation with one relaxation time. The maximum dielectric loss (maximum tan δ) is reached when the frequency of the EmF is the same as the c h a r a c t e r i s t i c   r e l a x a t i o n   f r e q u e n c y $\omega_x = 1/\tau$.

EmF energy can be applied to a physical object with a maximum linear dimension l in two ways (depending on the value of the ratio l/λ): The object can either form the load in an element of concentrated capacitance or inductance in the EmF generator circuit or it can be acted on by electromagnetic waves.

If the q u a s i - s t a t i o n - a r i t y   c o n d i t i o n $l \ll \lambda$ is fulfilled, the effect on the object can be estimated from direct-current laws. We consider this for the case of a semi-

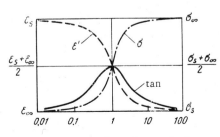

Fig. 1. Frequency dependence of ε', tan δ, and σ.

conducting cylinder with generatrix l, greater than the radius R, and with area of circular section S.

Let this object be placed in the electric field between capacitor plates so that the ends of the cylinder are parallel to the plates and equidistant from them. Then the current induced in the cylinder will be given by the relationship

$$I = \frac{V_{cyl}}{Z},$$ (14)

where V_{cyl} is the voltage drop between the ends of the cylinder, and $Z = l/S(\sigma - j\omega\varepsilon' \varepsilon_0)$ is its impedance.

The voltage drop on the cylinder is only a fraction of the voltage V_0 applied to the capacitor plates:

$$V_{cyl} = V_0 - \frac{2d}{S\omega\varepsilon_0} I,$$ (15)

where d is the distance from the end of the cylinder to the capacitor plate.

Let the cylinder now be placed in the magnetic field H inside a solenoid so that the axis of the solenoid coincides with that of the cylinder. This will give rise to a rotational induction electric field E in the cylinder:

$$E = \frac{\mu'\mu_0\omega RH}{2};$$ (16)

This field, in its turn, will produce eddy currents with density i:

$$i = E(\sigma - j\omega\varepsilon'\varepsilon_0).$$ (17)

In this case the current density is not uniformly distributed over the cross section of the cylinder, but decreases from the surface to the axis: The distance from the surface at which the current density is reduced by a factor e (= 2.71) is called the d e p t h o f p e n e t r a t i o n; this depth is calculated from the relationship

$$d = \frac{1}{\sqrt{\frac{\omega\sigma\mu'\mu_0}{2}}}.$$ (18)

If the maximum linear dimensions of the object are comparable with λ or exceed it, the power flux P_0 of the electromagnetic waves must be considered; it is partially reflected from the surface with a r e f l e c t i o n c o e f f i c i e n t equal to K:

$$K = \frac{(\sqrt{\varepsilon^*} - 1)^2}{(\sqrt{\varepsilon^*} + 1)^2},$$ (19)

and the rest is absorbed with increasing penetration and at a distance x from the surface is given by the relationship

$$P_x = P_1 e^{-2\alpha x},\qquad(20)$$

where $P_1 = P_0 (1-K)$ is the power absorbed in the object and
$\alpha = \omega \sqrt{\frac{\varepsilon}{2}(\sqrt{1+\tan\delta}-1)}$ is the absorption coefficient.

    In the general case the problem of absorption of wave energy
in an object is much more complex. For instance, recently Anne
et al. (1961) calculated the relative absorption cross section S —
the ratio of the power absorbed in a semiconducting sphere (σ, ε')
of radius R to the power incident on its cross section in the case of
propagation of a plane wave in air. If R > λ, then S = 0.5 ± 0.1, i.e.,
in this case about 50% of the power incident on the cross section of
the sphere is absorbed, irrespective of the value of λ and the value
of σ of the sphere material. If R < λ, however, the value of S for
particular values of R/λ (depending on σ and ε' of the sphere) may
be much greater than unity. These relationships are illustrated by
the graph shown in Fig. 2.

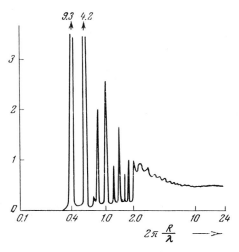

Fig. 2. Absorption of energy of a plane wave
(λ = 10.4 cm) in a semiconducting sphere
(ε' = 60, σ = 0.1 S/m) in relation to ratio of
radius R of sphere to wavelength λ.

Readers who wish to acquaint themselves more thoroughly with these questions are referred to the special literature (Kalashnikov, 1956; Tamm, 1957; Landau and Lifshits, 1957; Ramo and Whinnery, 1944).

Chapter 2

NATURAL AND ARTIFICIAL SOURCES OF ELECTROMAGNETIC FIELDS IN THE HABITATS OF ORGANISMS

2.1. The Earth's Electric Field

In the earth's atmosphere there is an electric field (E_e) in a direction normal to the earth's surface so that this surface is negatively charged and the upper atmosphere is positively charged. The strength of this field depends on the geographical latitude: It is greatest in central latitudes and decreases towards the equator and the poles. The accepted average value over the earth's sphere is $E_e = 130$ V/m. With increase in distance from the Earth's surface E_e decreases approximately exponentially and is about 5 V/m at a height of 9 km.

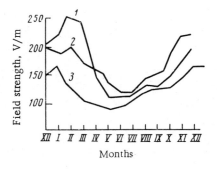

Fig. 3. Annual variation of electric field in atmosphere. 1) In Pavlovsk; 2) in Vysokaya Dubrava; 3) in Tashkent.

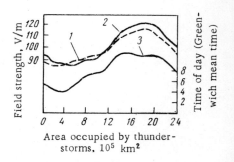

Fig. 4. Diurnal variations of electric field of atmosphere: 1) Over the ocean; 2) in polar regions; 3) thunderstorm activity over whole earth's surface.

The value of E_e undergoes periodic annual and diurnal variations. The annual variations, as Fig. 3 (Averkiev, 1960; Tverskoi,
1962) shows, are similar in nature over the whole earth's sphere
and have a maximum in December−February and a minimum in
May−July. The diurnal variations, however, have a planetary and
also a local nature. Over oceanic regions in various latitudes and
in polar regions the diurnal variation of E_e conforms to a single
universal law and is called the u n i t a r y   v a r i a t i o n. As the
graphs in Fig. 4 (Imyanitov and Chubarina, 1961) show, this variation is correlated with the total thunderstorm activity over the
earth's sphere, which undergoes the same diurnal variations. Over
the other land regions the diurnal variation E_e is also correlated
with the local thunderstorm activity and can vary considerably according to the time of the year.

2.2 T h e   E a r t h ' s   M a g n e t i c   F i e l d

The earth's magnetic field is distributed as shown in Fig. 5A.
This field is usually characterized by four parameters (Fig. 5B) −
the horizontal component of the intensity (H), the vertical component Z, the inclination angle I, and the declination angle D. The value of H is a maximum (0.3-0.4 Oe) at the magnetic equator and decreases to hundredths of an oersted at the poles; the vertical intensity Z decreases from 0.6-0.7 Oe at the poles and almost to zero at
the equator. In the regions of magnetic anomalies the values of Z
can be much higher than in neighboring regions (for instance, in the
region of the Kursk magnetic anomaly Z = 1.0-1.5 Oe). Recent discoveries are the regions of negative magnetic anomalies, where the
value of Z may be 0.02 Oe less than the normal value for the particular geographic latitude.

The earth's magnetic elements undergo variations with time,
i.e., the m a g n e t i c   a c t i v i t y varies. These variations are measured in gammas ($\gamma = 10^{-5}$ Oe) and are expressed either in terms
of a K index from 0 to 9 (corresponding to a change in the average
intensity from 4 to 500 γ) or in terms of the u measure, which is
calculated from the formula

$$u = \frac{0.1\,\Delta H}{\cos\Phi\cos(\psi - D)},\qquad(21)$$

where ΔH is the mean change of H in gammas, Φ is the geomagnetic ·
latitude, ψ is the angle between the geomagnetic and geographical

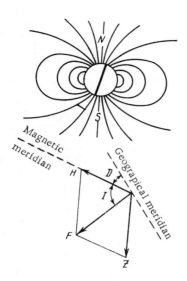

Fig. 5. A) Lines of force of earth's magnetic field; B) components of magnetic field. H) Horizontal components; Z) vertical component; D) declination angle; I) inclination angle.

meridians, and D is the declination angle.

Variations which appear at first sight to be random are called magnetic disturbances or (in the case of large variations) magnetic storms. These disturbances are of three kinds: synphasic — disturbances which occur sporadically and simultaneously over the whole planet; local — disturbances restricted to a particular region of the earth's surface, and permanent — disturbances which are continuously observed in some regions of the earth's surface. In synphasic and local magnetic storms the horizontal intensity increases the most — to several thousand gammas. While the synphasic variations differ only in maximum amplitude from place to place, the local variations differ in phase too. Permanent variations — up to hundreds of gammas — are observed continuously throughout the day, irrespective of the total magnetic activity.

All these kinds of magnetic activity are the result of solar activity associated with an increase in the number of sunspots and solar flares. Hence, the variations in magnetic activity are of a corresponding periodic nature. Firstly, there is a distinct 11-year cycle in the increase in magnetic activity coinciding with the years

Fig. 6. Relationship between mean annual magnetic activity U (curve 1) and number of sunspots R (curve 2).

of maximum sunspots, as Fig. 6 shows. Secondly, there is an annual periodicity with maxima of magnetic activity in the epochs of
equinox and minima in the epochs of solstice, and the amplitudes
of the variations depend on the solar activity (Fig. 8). Many weak
magnetic storms recur at 27-day intervals, a sequence which is
regularly renewed every 6-12 months. This is particularly notice-
able in the years of solar activity (when there are no bright flares).
This periodicity is correlated with the corresponding solar rotational period (Yanovskii, 1964; Ellison, 1955).

The diurnal periodic variation of the earth's magnetic elements is also associated with the sun. The amplitudes of these
variations change from day to day, but their phases remain constant. The greatest difference between the minimum and maximum
change in each element is observed in the spring—summer period
and the least difference is observed in the autumn—winter period.
Differences in the diurnal periodic variation in relation to geographical latitude are quite distinct. Figure 7 shows graphs of the diurnal periodic variations of the earth's magnetic elements in southern England (Ellison, 1955) and in Pavlovsk (Yanovskii, 1964).

Finally, there is a group of magnetic disturbance of a period-
ic nature which are called s h o r t - p e r i o d   f l u c t u a t i o n s (or
micropulsations of the magnetic field). The periods of these fluc-
turations range from hundredths of a second to several minutes,
but the amplitudes of the variations do not exceed a few gammas.
Thus, the total frequency spectrum of periodic variations of the
geomagnetic field covers the range from 10^{-5} to hundreds of hertz.

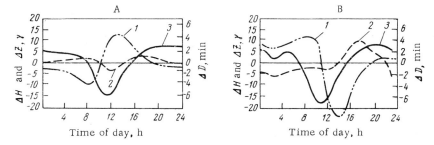

Fig. 7. Mean annual diurnal variation of earth's magnetic elements on "quiet" days
in southern England (A) and Pavlovsk (B). 1) ΔD; 2) ΔZ; 3) ΔH.

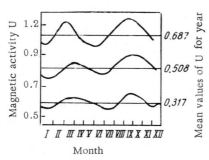

Fig. 8. Mean diurnal magnetic activity in years of strong (a), medium (b), and weak (c) disturbances.

2.3. Atmospherics

Atmospherics are EmFs created by atmospheric discharges, particularly lightning. The frequency range of atmospherics is very large — from hundreds of hertz to tens of megahertz. Their intensity is greatest at frequencies close to 10 kHz and decreases with increase in frequency (Fig. 9). In regions close to the locations of lightning discharges the strengths of the electric component of EmFs of atmospherics is of the order of tens, hundreds, and even thousands of W/m at frequencies close to 10 kHz.

The main centers of atmospherics are the continents in the tropics — regions of Central Africa and the central part of South America, southeastern USA (Florida), and southeastern Asia. The intensity of thunderstorm activity decreases towards high latitudes.

The intensity of thunderstorm activity (estimated from the area occupied by thunderstorms) varies with a diurnal periodicity. Figure 10 shows graphs of the diurnal variation of thunderstorm activity for the whole earth and for particular regions (Tverskoi, 1962). These graphs show that the minimum thunderstorm activity always and everywhere occurs in the morning hours and there is a increase at night. In the cold part of the year the maximum occurs in the middle of the night and in the warm part of the year it occurs at 3 to 6 p.m.

Thunderstorm activity also shows a seasonal periodicity. For instance, in the northern latitudes of the northern hemisphere the greatest number of thunderstorms occurs in summer (June—July) and the smallest number in winter.

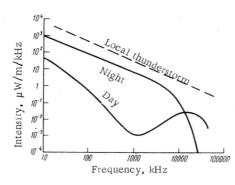

Fig. 9. Mean intensity of atmospherics in relation to frequency.

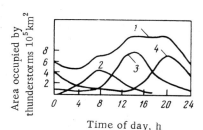

Fig. 10. Diurnal variations of thunderstorm activity over whole earth's surface (1), Asia and Australia (2), Africa and Europe (3), and America (4).

Finally, thunderstorm activity is correlated with solar activity: During solar flares (which are associated with sunspots) atmospheric are greatly increased.

2.4. Radio Emission of Sun and Galaxies

The radio emission of the sun and galaxies was first discovered about 20 years ago. It has now been established that the frequency range of these emissions is fairly broad — from 10 MHz to 10 GHz (Shklovskii, 1953; Chechik, 1953; Ellison, 1955; Handbook of Geophysics, 1965).

Figure 11 shows the distribution of the intensity of solar radio emission in this range in periods of the "quiet" sun and the "bursts" of emission in periods of the "disturbed" sun.

The flux of radio emission from the galaxies (from "radio stars") at a frequency of 100 MHz is of the order of 10^{-16} W/m^2/MHz.

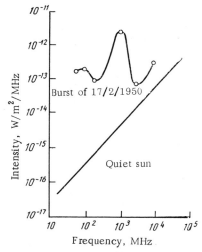

Fig. 11. Distribution of intensity of solar emission in the radio-frequency range.

The intensity of these radio emissions varies with a diurnal periodicity correlated with the rotation of the earth relative to the radiation sources. The intensity reaches its maximum in the morning and its minimum at night (Polukhanov, 1965).

In addition, the radio emissions vary in intensity with a periodicity of 27-28 days, correlated with the rotation of the sun and, finally, with the 11-year periodicity of solar activity.

2.5. EmFs in the Vicinity of Generators
of Various Frequency Ranges

The developement of electrical power engineering and radio and television techniques has led to the appearance of a great number of diverse EmF sources.

We will not give a full review of all the types of generators, but will merely mention the main parameters of those which are usually encountered in industrial enterprises, radio and television stations, and in medical institutions.

We must point out first of all the electromagnetic fields in the low-to ultrahigh-frequency range in the vicinity of generators must be regarded as induction fields, and not as fluxes of radio waves. As already mentioned, induction fields rapidly become weaker with increasing distance from the source and beyond a circle of radius equal to a few wavelengths (where the work points of the servicing personnel are usually situated) the EmF strengths are a very small fraction of their initial values.

EmFs of industrial frequency (50 Hz) are produced near electric transmission lines, transformers, and so on. In the immediate vicinity of these sources the EmF strengths can be very considerable. For instance, in the case of high-voltage (400 and 500 kV) open distributing devices the servicing personnel may be in a field with an electric component of the order of several thousand V/m (Sazonova, 1964; Lebedeva, 1963).

High-frequency EmFs — with frequencies from tens to hundreds of kilohertz — are most intense close to industrial generators for high-frequency tempering of metals, drying of timber, and so on. In these conditions E can attain values of thousands of V/m and H tens of A/m at work points (Nikonova, 1963).

Ultrahigh-frequency EmFs — from a few MHz to tens of MHz — are most intense in the workrooms of radio and television stations, where the field strength E can reach hundreds of V/m (Fukalova, 1964a).

Superhigh-frequency EmFs — from hundreds to thousands of MHz, which are produced close to the corresponding installations (radar installations, for instance), can be evaluated from the power flux density, which may have values of several mW/cm^2.

2.6. The "Radio Background"

A "radio background" is created around the earth by the transmission of the numerous radio and television stations. It is very difficult to estimate the intensity of the "radio background" and its variation with time. However, some general points can be made.

Firstly, we must distinguish between the regions situated in the vicinity of radio and television stations and the regions remote from such stations. In regions of the first type the intensity of the "radio background" may be fairly high — of the order of tenths of a V/m. In the remote regions the intensity of the "radio background" is much lower and the main contribution to it is made by short-wave stations. Since all the stations transmit incoherently, the "radio background" is the result of summation of the transmissions.

Variation in the intensity of the "radio background" in relation to the time of day occurs only in the regions of the first type, where the main sources of "radio background" are long-wave and medium-wave stations, and also television stations, which operate in the meter range. These stations usually cease to operate in the period between 1 a.m. and 6 a.m., approximately. Short-wave stations, however, which transmit over the whole earth, operate practically around the clock.

A general idea of the intensity of the "radio background" can be obtained by comparing it with the level of atmospheric interference. It is believed that the intensity of radio signals is 10-100 times greater than that of interference (this ensures reliable reception of the radio signals).

Chapter 3

ELECTRIC PROPERTIES OF THE TISSUES OF LIVING ORGANISMS

Theoretical and experimental investigations of the electric properties of the tissues of living organisms have been the subject of numerous papers, including several reviews (Schwan, 1948, 1954, 1955, 1957, 1959; Hartmuth, 1954; Presman, 1956a, 1964a, 1964b, 1965a).

As regards electric properties the tissues of living organisms can be divided into three groups according to their water content: suspensions of cells and protein molecules of liquid consistency (blood, lymph), similar suspensions in a condensed state (muscle, skin, liver, etc.), and tissues with a low water content (fat, bone). Cells, colloidal particles, protein molecules, and other micropart- icles acquire a dipole moment when they are suspended in an elec- trolyte solution. Electric charges in tissues are also represented by the dipole molecules of water and, finally, by the electrolyte ions.

3.1. Properties of Tissues in Constant Fields

Tissues in a constant electric field are polarized to some extent — the charged particles move along the lines of force and the dipole molecules are oriented in the same direction. If a con- stant voltage is applied directly to tissue an electric current, due to ionic conduction, will be produced in it.

Each cell is surrounded by a wall with a surface capacitance of 0.1–3 μF/cm and a surface resistance of 25 to 10,000 $\Omega \cdot$ cm^2. The intercellular and intracellular media have a resistance of the order of 100–300 $\Omega \cdot$ cm and a dielectric constant of about 80. Fig- ure 12 shows the equivalent circuit of a cell and the extracellular medium.

It is obvious that on application of a constant voltage the cell wall will act as an insulator and the current can flow only in the extracellular medium. A constant voltage can also give rise to the phenomenon of electrophoresis — the migration of electrically charged particles (cells, macromolecules).

3.2 Dispersion of Electric Parameters of Tissues in Alternating Fields

The electric properties of living tissues in alternating EmFs have been dealt with most thoroughly in the works of Schwan and his colleagues (Schwan, 1953b, 1956; Schwan and Li, 1953; Schwan and Kay, 1956; Schwan and Carstensen, 1957; Rajewsky and Schwan, 1948; etc.), and several other researchers (Cook, 1951a, 1951b, 1952b; Cole and Cole, 1941; and others). We will discuss only some

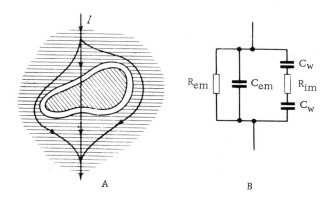

Fig. 12. Passage of electric current (I) in cell (A) and equivalent electric circuit of cell (B). R_{em} is the resistance and
C_{em} is the capacitance of the extracellular medium. R_{im} is
the resistance of the intracellular medium. C_w is the capacitance of the cell wall.

of the main conclusions derived from these investigations.

In the region of EmFs of interest to us — from infralow to
superhigh frequencies — there are three frequency ranges in which
ε' and σ (or $\rho = 1/\sigma$) of tissues vary with the frequency: the α-dis-
persion range at low frequencies, the β-dispersion range at radio
frequencies, and the γ-dispersion range at superhigh frequencies.
Figure 13 shows a graph illustrating the course of these three kinds
of dispersion for muscle tissue (of man and other mammals). A
similar dispersion curve is obtained for other tissues with a high

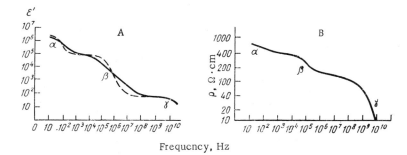

Fig. 13. Dielectric constant (A) and resistivity (B) of muscle tissue in relation to frequency. The broken line is the theoretical curve.

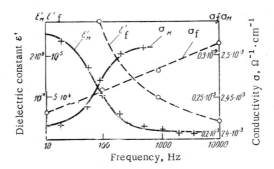

Fig. 14. Dielectric constant and conductivity
of muscular (ε'_m, σ_m) and fatty (ε'_f, σ_f) tissues at low frequencies. The continuous curves
are calculated from the Debye formula (12).
The crosses and circles are the experimental
data.

water content. For tissues with a low water content, such as fatty
tissues, the curve is of the same general nature, but the values of
ε' and σ are approximately an order lower.

Figure 14 shows the variation of ε' and σ of tissues at low
frequencies (α-dispersion). Schwan (1957) expresses the following
hypotheses on this kind of dispersion:

1. Since only ionic conduction is possible at low frequency
and the cell walls behave as insulating layers, the low-frequency
currents can flow only in the extracellular medium and this accounts for the low conductivity of the tissues. Fatty tissues in themselves have a low conductivity and their electrolyte content is very
low. The increase of σ with frequency increase can be attributed
to the corresponding reduction in the capacitive reactance of the
cell wall, which leads to an ever-increasing implication of the intracellular medium in the total conductivity of the tissue.

2. The very high values of ε' at low frequencies and the sharp
decrease in the values with frequency increase are due to the relaxation of charging and discharging processes on the cell wall by the
ionic atmosphere surrounding the electrically charged surface off
the cell. The frequencies are so low that the cell walls become
charged (owing to ions outside and inside the cell) in one cycle.

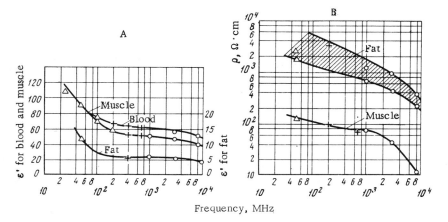

Fig. 15. Dispersion of dielectric constant (A) and resistivity (B) of blood, muscle, and fat in the radio-frequency range.

Hence, the total charge during the cycle is high and the capacitance of the tissue is considerable. This is equivalent to a high dielectric constant of the tissue (capacitance per unit volume).

The β-dispersion in the region of higher frequencies, right up to superhigh frequencies, is shown graphically in Fig. 15 (Schwan and Piersol, 1954; Schwan, 1957). With frequency increase ε' decreases until the cycle becomes so short that the cell wall cannot become charged (in other words, the capacitive reactance of the wall becomes very low). In the case of blood such conditions occur at a frequency of about 100 MHz. The reduction of the resistivity up to this frequency is due to the reduction in the capacitive reactance of the cell wall with the result that the intracellular contents play an increasing part in the total conductivity of the tissues. Investigations of the electric properties of blood have shown (Schwan, 1953a, 1955; Cook, 1951b; and others) that the dispersion of this type is described satisfactorily by equation (12) with the following parameters:

$$\tau = RC_m \frac{\sigma_i + 2\sigma_a}{2\sigma_i \sigma_a + RG_m(\sigma_i + 2\sigma_a)},$$

$$\varepsilon'_s = \frac{9}{4\varepsilon_0} \frac{pRC_m}{\left[1 + RG_m\left(\frac{1}{\sigma_i} + \frac{1}{2\sigma_a}\right)\right]^2},$$

(22)

$$\sigma_\infty = \sigma_a \left(1 + 3p \, \frac{\sigma_i + \sigma_a}{\sigma_i + 2\sigma_a} \right),$$

$$\sigma_s = \sigma_a \left[1 + \frac{3}{2} \, p \, \frac{1 + RG_m\left(\frac{1}{\sigma_i} - \frac{1}{\sigma_a} \right)}{1 + RG_m\left(\frac{1}{\sigma_i} + \frac{1}{\sigma_a} \right)} \right], \tag{23}$$

where R is the radius of the cell, p is the fraction of the volume occupied by the cells, and C_m and G_m are the capacitance and conductivity of the cell membrane. The subscripts i and *a* refer to the intracellular and extracellular contents, respectively.

At frequencies of the order of several MHz this mechanism of dispersion, due to cell-wall relaxation, is supplemented by the comparatively weaker effects of relaxation of dipole protein molecules (characteristic frequencies close to 10 MHz) and the structural relaxation of subcellular components (Schwan, 1959).

It is difficult to explain the small dispersion for blood in the frequency range from 100 MHz to GHz. At these frequencies the cell walls cease to have any effect (they become "short-circuited") and the relaxation of the polar water molecules still does not occur. The ion content of the electrolyte medium of tissues has no effect on the dispersion of the dielectric constant and, hence, in the considered frequency range ε' of an electrolyte medium is practically independent of the frequency. The same is true for the conductivity of electrolytes.

It has been suggested (Schwan, 1957, 1959) that in this frequency range the electric properties of tissues with a high water content are due to an electrolytic medium containing suspended protein molecules with a lower value of ε'. In other words, the macromolecules form "dielectric cavities" in the electrolyte. The dielectric constant of such a macromolecular suspension can be determined from Fricke's equation (1925):

$$\frac{\varepsilon_\infty - \varepsilon_a}{\varepsilon_\infty + \chi \varepsilon_i} = p \, \frac{\varepsilon_i - \varepsilon_a}{\varepsilon_i + \chi s_a}, \tag{24}$$

where χ is a factor allowing for the ellipsoidal shape of the molecules.

This equation can be used to calculate the "effective" dielectric constant of a hydrated protein, regarded together with the hydrate sheath as a unit whole.

The dispersion of ε' of such tissues as blood (or other tissues with a high water content) can be attributed to the variation of the effective dielectric constant of protein molecules with frequency. Attempts have been made to attribute this relationship, found for hemoglobin molecules, to two possible mechanisms (Schwan, 1959).

1. The protein molecule contains relatively free groups capable of vibrating under the action of EmFs with a frequency greater than the characteristic frequency for rotation of the molecules as a whole. A spectrum of characteristic frequencies in the considered range has actually been found for some amino acids and peptides (Schwan, 1955).

2. The dielectric properties of the water envelope exhibit dispersion at frequencies above 100 MHz. It has been suggested that as regards dielectric properties "bound" (hydrating) water occupies an intermediate position between ice (the characteristic frequency of which lies in the range of audio frequencies) and free water (about 2 GHz). The characteristic frequencies of bound water lie in the intermediate range.

The situation is more difficult as regards the dispersion of σ in this range. The calculated data obtained from a formula similar to (24) differ considerably from the experimental values. This is presumably due to the concentration-dependent effect of the ionic bonds of the protein component of the macromolecular suspension.

The nature of the γ dispersion at frequencies above 1 GHz can be satisfactorily accounted for by the polar properties of water molecules. The dispersion curves agree fairly well with the Debye equations (12) if a term which takes the ionic conduction into account is introduced into the expression for ε'' (Cook, 1951),

$$\varepsilon'' = \frac{\varepsilon'_s - \varepsilon'_\infty}{1 + (\omega\tau)^2}\omega\tau + 1.3 \cdot 10^{13}\frac{\sigma}{\omega\varepsilon}, \tag{25}$$

where τ is the relaxation time for water molecules (of the order of 10^{-11} sec), and σ is the ionic conductivity which is independent of the frequency.

The nature of the dispersion for fatty tissues depends on their structure. It has been found (Cook, 1951) that the parameters of purely fatty tissues are practically independent of the frequency in the range above 100 MHz, whereas dispersion is observed in tissues

Table 2. Values of Electric Parameters of Animal and

Fre-quency	Muscle	Heart muscle	Liver	Lungs	Spleen	Kidneys
					Resistivity, ρ, $\Omega \cdot$ cm	
10 Hz	[a]960	[a]960	[a]840, 1220	[a]1100	—	—
100 Hz	[a]890	[a]920	[a]800, 1060	[a]1110	—	—
1 kHz	[a]800 980	[a]750, 930 [e]830—900 [g]700—1300	[a]770, 800,970 [e]1000—1600	[a]1000 [e]1400—1900	[a]1000 [g]260—430	—
10 kHz	[a]760, 880	[a]600	[a]685, 860	[a]900	—	—
100 kHz	[i]170—250 [f]520	[i]190—240	[a]460 [i]220—550 [g]550—800 [j]420	[i]165—200	[i]250—500	[i]150—270
1 MHz	[i]160—210 [f]250	[i]180—230 [g]400—550 [j]400	[i]210—420	[i]150—280	[i]200—380	[i]140—250
10 MHz	[i]150—170	140—180	[i]180—260	[i]110—150	[i]150—170	[i]120—170
100 MHz	[k]100—130 [i]120—160 [m]140—200 [m]120—150	[i]130—170	[k]120—145 [i]150—200 [m]180—210 [m]150—180	[k]95—130 [i]100—140	[k]85—105 110—150 [m]150 [m]120	[k]100—120 100—150 [m]130—160 [m]90—140
1 GHz	[n]75—79 [o]81—84 [p]77	[o]83—100	[n]98—106 92—100 [P]100	[o]137	—	[o]81—82
10 GHz	[n]12 [q]13	—	[n]15—17	—	—	—
24 GHz	—	—	—	—	—	—

Human Tissues in Various Frequency Ranges

Brain	Fatty tissue	Bone	Bone marrow	Blood	Plasma	0.9% NaCl solution
—	—	—	—	—	—	—
—	—	—	—	b 166	—	—
e 500—800 g 450—550	a —500—5000 c 1700—2500			b 166 d 147 e 120—135 h 130—180	b 60	—
—	—	—	—	d 147	—	—
i 460—880	—	—	—	d 147	—	—
i 430—700	—	—	—	d 140	—	—
i 300—450	—	—	—	d 90	—	—
k 160—230 i 200—300 m 220—260 i 180—200	k 1170—1250 i 1500 m 2200—4300 m 1700—2500	—	m 4100—5300 m 3000—5000	l 82 m 120—150 m 80—100	k 61 l 70 m 80 m 60	60
—	700—1400 1100—3500 P2700	2000	n 1000—2300	n 64—72 o 80	n 54	49 56 53
—	n 240—370 q 210	n150 q 130	n 60—200 q 100	q 11 q 9,5 r 9,3	n 9	9
—	t 71	t 71	—	t 3,8	—	—

Table 2. (Continued)

Fre-quency	Muscle	Heart muscle	Liver	Lungs	Spleen	Kidneys
						Dielectric constant,
10 Hz	a $10000 \cdot 10^3$	a $7000 \cdot 10^3$	a $16000 \cdot 10^3$	a $8000 \cdot 10^3$	—	—
100 Hz	a $800 \cdot 10^3$ s $1000 \cdot 10^3$	a $800 \cdot 10^3$ a $820 \cdot 10^2$	a $900 \cdot 10^3$ a $850 \cdot 10^3$	a $450 \cdot 10^3$	—	—
1 kHz	a $130 \cdot 10^3$ f $100 \cdot 10^3$ s $170 \cdot 10^3$	a $300 \cdot 10^3$	a $150 \cdot 10^3$	a $90 \cdot 10^3$	—	—
10 kHz	a $50 \cdot 10^3$ a $60 \cdot 10^3$ $90 \cdot 10^3$ f $50 \cdot 10^3$	$100 \cdot 10^3$	a $50 \cdot 10^3$ a $60 \cdot 10^3$	a $30 \cdot 10^3$		
100 kHz	f $20 \cdot 10^3$ s $30 \cdot 10^3$		g 7000—12000	—	—	—
1 MHz	f 2000	—	g 1200—2000	—	—	—
10 MHz	—	—	—	—	—	—
100 MHz	m 69—73 m 71—76	—	k 65—75 m 72—74 m 70—79	—	m 38—90 m 100—101	m 83—84 m 87—92
1 GHz	n 49—52 o 53—55 p 61	o 53—57	n 46—47 o 44—52 p 50	.35	—	o 53—56
10 GHz	n 40—42 q 29		n 34—38	—	—	—
24 GHz	—	—	—	—	—	—

Notes. a) Dog, in situ, at body temperature; b) sheep, at 18°C (plasma at 37°C): c) dog, in situ, at body temperature; d) rabbit, at 20°C; e) rabbit, fresh tissue, at 23°C; f) rabbit, fresh tissue, at 20°C; g) man and various animals, fresh tissues at 23°C; h) sheep, at 18°C; i) man, minced tissue, at 18°C; j) rabbit, minced tissue, at 23°C; k) man, minced tissue, at 37°C; l) sheep, at 20°C; m) ox and pig, fresh tissue, at 20°C; n) dog and horse, fresh tissue and blood, at 38°C; bone and bone marrow at 25°C; o) man, fresh tissue, at 23°C; p) ox, minced tissue, at 22°C; q) man, fresh tissue, at 37°C; r) man, fresh tissue, at 35°C; s) frog, fresh tissue, at 25°C; t) man, fresh tissue, at 37°C.

Brain	Fatty tissue	Bone	Bone marrow	Blood	Plasma	0.9% NaCl solution
relative to vacuum						
—	—	—	—	—	—	—
—	$a_{150}\cdot10^3$	—	—	—	—	—
—	$a_{50}\cdot10^3$	—	—	—	—	—
—	$a_{20}\cdot10^7$	—	—	d 2900 d 2800	—	—
—	—	—	—	d 2740	—	—
—	—	—	—	d 2040	—	—
—	—	—	—	d 200		
m 70—75 m 81—83	m 8—13 m 11—13	—	m 7—8	m 72—74 m 73—76	m 82 m 76	78
—	n 4,3—7,5 o 3,2—6 p 9,5	n 8	n 4,3—7,8	n 58—62 63	n 69	72 78
—	n 3,5—3,9 q 3,6	n 8 q 6,6	n 4,4—6,6 5,8	n 50—52 45 48	n 61	66
—	t 3,4	t 6,3	—	t 3,2	—	—

consisting of fat cells surrounded by an electrolytic medium. For bony tissues the dispersion satisfies the Debye equations with a relaxation time of $0.7 \cdot 10^{-11}$ sec and a correction for ionic conductivity.

3.3. Values of Electric Parameters of Tissues

Measurements of the electric parameters of living tissues have been made by the methods usually employed in electric and radio-engineering measurements (Course in Electrical Measurements, 1960; Brandt, 1963). Impedance bridges have been used at low frequencies (Schwan, 1953b; Schwan and Sittel, 1953a, 1953b), resonance methods employing twin cables (Schwan, 1950, 1955; Schwan and Li, 1955a, 1955b; and others) and coaxial cables (Laird and Ferguson, 1949; Cook, 1951a; and others) in the radio-frequency range, and waveguide lines at superhigh frequencies (England and Sharples, 1949; Herrick et al., 1950; Jackson, 1946; Grouch, 1948; Livenson, 1964a; and others). Waveguide bridges have been used for measurements on liquid substances at superhigh frequencies (Buchanan, 1952). Finally, measurements of the electric parameters of small samples of biological materials at superhigh frequencies have been made by means of cavity resonators (Shaw and Windle, 1950; Bayley, 1951).*

The measurements have usually been made on fresh samples of tissues and uncoagulated blood taken from man and various animals and kept at constant temperature in the range 18–38°C during the measurements. Measurements have also been made in situ.

Table 2 gives the values of the resistivity and dielectric constant of various tissues at frequencies of 10 Hz to 24 GHz. The table is compiled from data taken from the "Handbook of Biological Data"(1956) and from other sources (Roberts and Cook, 1952; Schwan, 1957; Cook, 1951a; England and Sharples, 1949; England, 1950; Herrick et al., 1950; Horn, 1965). The values of ε' and ρ vary with the temperature. If their frequency dependence is weak, these changes are

$$\frac{\Delta\rho}{\rho} = -2 \ \% \ /°C \ \text{and} \ \frac{\Delta\varepsilon'}{\varepsilon'} = -0.5 \ \% \ /°C.$$

---

* Waveguide components and cavity resonators will be discussed in Section 5.3.

Chapter 4

PHYSICAL PRINCIPLES OF THE
INTERACTION OF ELECTROMAGNETIC
FIELDS WITH BIOLOGICAL OBJECTS

As already mentioned in the introduction, attempts to determine the physical mechanisms of the biological action of EmFs and to estimate the minimum intensity at which EmFs can act on biological objects have usually been based on a consideration of the energetic interactions of EmFs with these objects. In some cases the center of interest has been the physicochemical processes induced in living tissues by EmFs and in others it has been the relationship between this action and the macroscopic parameters of the biological objects.

4.1. Biological Objects in an
Electrostatic Field

The tissues of living organisms situated in an electric field have electric charges induced on the interfaces of media with different electric parameters and the polarization of bound charges also takes place.

Most tissues in an electrostatic field can be regarded as conducting media and the body of a man or animal can be regarded in a first approximation as a homogeneous conductor. On this assumption we can estimate the distribution of the charge induced on the body surface from the formulas derived for conducting bodies of simple geometric shapes in an electric field (Stratton, 1941; Landau and Lifshits, 1957). For instance, the human body can be regarded as a homogeneous conducting ellipsoid. If such an ellipsoid is situated in a uniform electrostatic field of strength E and its major axis is parallel to the field lines, the density of the induced surface charge is given by the relationship

$$q_{sur} = \frac{E}{4\pi n} \cos \theta, \tag{26}$$

where n is a coefficient which depends only on the form of the ellipsoid and θ is the angle between the direction to the considered point of the surface and the direction of the field.

In this case the distribution of surface charges is such that the ellipsoid acquires a dipole moment p (along the major axis) equal to

$$p = \frac{abcE}{3n},$$ (27)

where a, b, and c are the semiaxes of the ellipsoid.

For conducting bodies of spherical shape with radius a the corresponding formulas have the form

$$q_{\text{šur}} = \frac{3E}{4\pi} \cos\theta, \quad p = a^3 \cdot E.$$ (28)

Hill (1958) theoretically examined the possible mechanism of interaction of an electrostatic field with tissue macromolecules. The electric field causes polarization of the macromolecules in the solution due to the fact that the molecules possess a constant dipole moment and to the change in the position of protons in the molecule. Such action can affect the relative stability of the two possible configurations of the macromolecules. On the basis of these considerations the author concluded that the action of fields with strengths of the order of 10,000 V/cm can cause separation of DNA strands (transition from the paired to the unpaired state) and this can act as a trigger for separation of the chromosomes in the cell nucleus before cell division. Another possibility is the effect of the field on the state of the protein chains in muscle fibers (transition form the long to the short chain), which may act as a trigger for muscular contraction.

4.2. Biological Objects in a

Magnetostatic Field

A constant magnetic field can in principle affect various processes in biological objects: Up to twenty possible kinds of interaction have been listed (Barnothy, 1964). Within recent years there have been a few attempts at a theoretical treatment of the main physical mechanisms of the biological effects of a magnetic field and estimates of the field intensities which can produce such effects. These theoretical investigations can be divided into two main groups according to the kind of magnetic field effects (microscopic or macroscopic) which are considered.

In the first group of investigations the initial assumption is that the mechanisms of biomagnetic effects are due to physical phenomena induced at the molecular, and even atomic, level. For instance, some authors see the main cause of biomagnetic effects in the orientation of diamagnetic or paramagnetic molecules by the magnetic field. Others believe that this field can cause distortions of the valence angles in molecules. Others draw attention to the orientation of the spins of molecules in a magnetic field, and so on.

Dorfman (1962) points to the fact that protein macromolecules, which are usually diamagnetic and have axial symmetry, must exhibit fairly high magnetic anisotropy (difference in magnetic susceptibility χ along and across the axis of the molecule) of the order of $\Delta\chi = 10^{-22}$. Hence, in a protein solution situated in a uniform magnetic field of strength $H = 10^4$-10^5 Oe almost 100% orientation of the macromolecules may take place. In a nonuniform field with an intensity gradient $dH/dx = 10^4$ Oe/cm the effect of such orientation can lead to the creation of a corresponding gradient of concentration of the protein molecules in the solution. However, in very viscous biological solutions the establishment of orientation equilibrium for anisotropic molecules is not very likely. In addition, thermal motion will upset the orientation of the macromolecules due to the field if the magnetic energy, calculated for one molecule, does not exceed kT. This can happen only at very high field intensities — more than 10^5 Oe.

Yet other authors (Antonov and Plekhanov, 1960) have pointed to the fact that the cohesive forces between molecules lead to their short-range ordering, i.e., can facilitate the orientational effect of a magnetic field against a background of thermal motion; such an effect becomes probable at an intensity of only 100 Oe. Gross (1964) postulates that biomagnetic effects can be produced at relatively low field intensities (of the order of thousands of Oe). In his opinion, such effects can be produced by orientation of the magnetic moments of unpaired electrons in free radicals, irrespective of the disturbing effect of thermal motion. He also believes that there is another possible mechanism of the biomagnetic effect: distortion of the bond angles in paramagnetic molecules.

A quantum-mechanical model of the mechanism of action of a magnetic field has also been considered by Valentinuzzi (1964), who introduces a reducing factor for biochemical reactions. This factor is the ratio of the number of molecules at a low stable level (magnetic moment parallel to the field) to the number of molecules at a higher level (moment antiparallel to the field). According to Valentinuzzi's calculations, the reducing factor for reactions involving methemoglobin in a field of intensity $5 \cdot 10^3$ Oe will be 0.99, and at $8 \cdot 10^5$ Oe the reactions will be completely stopped.

It has recently been suggested (Neprimerov et al., 1966) that ortho to para transitions* may take place in water molecules in a magnetic field. The magnetic energy required for this (calculated per molecule) is very low — for instance, 100 times less than for the breakage of weak hydrogen bonds in a molecule. Ortho to para transitions in aqueous solutions can give rise to regions with parallel orientation of the spins and this will lead to the expulsion of dissolved substances from such regions.

The macroscopic mechanisms of biomagnetic effects have been investigated on various models. For instance, Neurath (1964) showed that in a magnetic field of intensity $3 \cdot 10^5$ Oe erythrocytes rotate at a rate of 68 deg/min, i.e., twice as fast as the thermal motion, but the establishment of the equilibrium state of such an effect is very slow. Neurath considers that a more probable effect is the production of an electric potential gradient in the blood vessels by the electric field (magnetoelectric effect). For instance, in the aorta with a blood flow speed of 100 cm/sec a magnetic field intensity of 500 Oe will induce an electric field with a gradient of 0.14 mV/cm, and at an intensity of $5 \cdot 10^5$ Oe the corresponding gradient will be 5 mV/cm, which is comparable with the sensitivity of nerve cells (10 mV/cm). Another example is the induction of an electric field with a gradient of 0.1 mV/cm in the blood vessels of fish when they are rotated at 180 deg per second in a magnetic field of intensity 500 Oe. This gradient greatly exceeds the sensitivity of the special receptors of fish, which react to gradients of the order of 10^{-5} mV/cm.

The mechanism of biomagnetic effects has been considered from the viewpoint of magnetochemical effects by Dorfman (1966). He considers the pulsating pressures which can be produced in tissues when a magnetic field interacts with biocurrents with frequencies varying from 10 to $2 \cdot 10^3$ pulses/sec. According to Dorfman's calculations, at intensities of 10^2–10^3 Oe pulsating ponderomotive

---

*Transitions of water molecules from the ortho state, where the spins of the protons of the two hydrogen atoms are parallel, to the para state, where the spins are antiparallel.

forces exerting a pressure of the order of 10^{-6} to 10^{-1} dyn/cm^2 can be produced in regions where the biocurrents flow. The sensitivity of the human ear (10^{-4} dyn/cm^2) lies in precisely these limits. It is an interesting suggestion of Dorfman that resonance effects of such kind can occur when the frequency of the induced mechanical vibrations in the particular region of the organism (or organ) coincides with the natural frequency of its free vibrations. In this case the magnetochemical effect may be significant even at very low field intensities, in the geomagnetic field, for instance.

Thus, most authors, proceeding from theoretical arguments and calculations based on microscopic and macroscopic concepts, conclude that biomagnetic effects are possible only at fairly high field intensities — at least thousands of Oe.

4.3. Absorption of EmF Energy in Tissues and Its Conversion to Heat

Investigations of the conversion of EmF energy to heat have been the subject of a considerable number of theoretical and experimental studies, since this mechanism of interaction of EmFs with living tissues has been regarded at the only possible cause of any of the biological effects produced by EmFs of high, ultrahigh, and superhigh frequencies for the treatment of various diseases. On the basis of thermal concept some investigators have tried to evaluate the maximum permissible intensities of radio-frequency EmFs from the viewpoint of occupational hazard.

As we mentioned in the introduction, the thermal concept of the biological effects of EmFs contradicts the results of several investigations carried out with EmFs of low intensity. However, in cases where biological objects are exposed to EmFs of fairly high intensities (at which the thermal effect is possible) this concept is useful. Hence, we will consider in detail the theoretical and experimental data on the thermal effects of EmFs of various frequencies.

In the low-frequency and high-frequency ranges the conversion of EmF energy to heat energy is due mainly to conduction losses resulting from the release of Joule heat in the tissues by the ionic currents induced in them.

Up to frequencies of the order of 10 MHz the dimensions of the body of man and large animals (and, of course, of small animals) are small in comparison with the wavelength and the body tissues can be regarded as a conducting medium. Hence, the quasi-stationarity conditions are satisfied and calculations can be made from formulas (26)-(28), which are derived for a static field. The EmF power absorbed in unit volume of the body can be calculated in this case from the direct-current laws:

$$P = i^2\rho \quad \text{W/cm}^3 \tag{29}$$

The current density i must be calculated for the particular shape and electric parameters of the biological object. Such a calculation has been made for a man situated in an alternating electric or magnetic field in the frequency range 100 kHz to 1 MHz on the following assumptions (Presman, 1957a, 1960a):

1. The human body is regarded as a homogeneous (in electric properties) conducting ellipsoid;

2. The only case considered is that of a uniform electric or magnetic field in which the body (ellipsoid) is placed so that its major axis is parallel to the lines of force (Fig. 16).

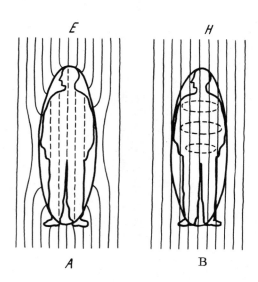

Fig. 16. Human body (ellipsoid) in a uniform electric (A) and magnetic (B) field. The broken lines indicate the direction of the induced current.

In these conditions the current density in the case of the electric field is

$$i_E = 1.3 \cdot 10^{-13} \cdot f \cdot E \quad \text{A/cm}^2 \tag{30}$$

and in the case of the magnetic field is

$$i_H = 1.3 \cdot 10^{-11} \cdot f \cdot H \quad \text{A/cm}^2 \tag{31}$$

(E is expressed in V/m, H in A/m, and f in Hz).

The amount of heat produced in the human body in this case is given by the relationship

$$Q_E = 2 \cdot 10^{-20} \cdot \rho_{av} \cdot f^2 \cdot E^2 \quad \text{cal/min} \tag{32}$$

$$Q_H = 2 \cdot 10^{-16} \rho_{av} \cdot f^2 \cdot H^2 \quad \text{cal/min} \tag{33}$$

(ρ_{av} is the average resistivity of the human body tissues, the value of which in the considered ranges can be taken as equal to that for muscle tissues; see Table 2).

In the ultrahigh- and superhigh-frequency ranges the conversion of EmF energy to heat energy entails not only conduction loss, but dielectric loss. The fraction of dielectric loss in the total EmF energy absorbed in the tissues increases with frequency. For instance, the loss due to relaxation of water molecules in tissues at a frequency of 1 GHz is about 50% of the total loss. At a frequency of 10 GHz it is about 90% and at a frequency of 30GHz about 98% (England, 1950).

In these frequency ranges (above 100 MHz) the dimensions of the body of man and large animals become comparable with λ or exceed it and the body tissues can no longer be regarded as a conducting medium. Finally, the tissues cannot be regarded as homogeneous in electric properties. In other words, the quasi-stationarity condition is not satisfied in this case and we are dealing with a wave flux, part of which is reflected from the body surface and the rest of which is gradually absorbed in electrically inhomogeneous tissues.

With allowance for reflection the EmF power absorbed per square centimeter of surface of the object, or the effective power (P_e), will be

$$P_e = P_0 \cdot (1 - K), \tag{34}$$

where P_0 is the power flux density incident on the surface of the object and K is the reflection coefficient.

The values of the reflection coefficient for EmFs of different frequencies for various tissues [calculated from formula (19)] are given in Table 3 and the depth of penetration of EmF energy into the tissues (i.e., the depth at which the energy is reduced by a factor e) is given in Table 4 (Fleming et al., 1961).

The relationship between the degree of absorption of EmF energy in a biological object and the dimensions of the latter can be determined from the previous calculations (see Fig. 2) for a semiconducting sphere. It follows from these calculations that when R > λ about 50% of the power incident on the cross section of a semiconducting sphere is absorbed in it, irrespective of the value of σ of the sphere material. Calculations (Anne et al., 1961) and experiments on models (Mermagen, 1961) have shown that this is true for biological objects of any shape in the frequency range from 300 MHz to 3 GHz. When R < λ, however, the power absorption depends on the electric parameters of the object and at certain values of R/λ more power is absorbed in it than is incident on the cross section.

The relationship between the absorption and the anatomical position of the tissues also depends mainly on the thickness of the subcutaneous fat and the method of applying the EmF to the object. If the object is exposed to the field by being placed between the plates of a capacitor, then in the subcutaneous layer, which has lower values of ε' and σ than in the deeper muscle tissues, the strength E will be greater than in the muscles. The absorbed EmF power will be distributed in a corresponding manner (Schwan and

TABLE 3. Coefficient of Reflection from Interfaces
between Tissues at Different Frequencies

Interfaces	Frequency, MHz							
	100	200	400	1000	3000	10 000	24 500	35 000
Air−skin	0.758	0.684	0.623	0 570	0.550	0.530	0.470	—
Skin−fat	0.340	0.227	—	0.231	0.190	0.230	0.220	—
Fat−muscle	0.355	0.3515	0.3004	0.2608	—	—	—	—

Piersol, 1954). If the object is exposed to waves the fatty layer may act as an "impedance transformer" between the air and the muscle tissue and this may lead to some compensation of the reflection of the waves and, hence, to a corresponding increase in the fraction of power absorbed. This effect depends on the thickness of the fatty layer, the thickness of the skin layer, and the frequency of the EmF. Figure 17 shows how the fraction of absorbed energy varies when these parameters vary (Schwan and Li, 1956b).

The percentage of EmF energy absorbed in various body tissues can be calculated approximately from the electric parameters of the tissues. Such a calculation has been carried out for the case of exposure of the occipital region of the head of man, rabbits, and rats to EmFs of superhigh frequency (Presman, 1965a). Figure 18 shows the frequency characteristics of the relative distribution of absorbed energy in different layers of the head tissue.

So far we have ignored one physical process which may affect the relative distribution of absorbed EmF energy in the tissues of living organisms, viz., the production of standing waves as a result of which the energy absorbed in a particular layer of tissue may be much greater than in the case of propagation of waves in this tissue. Standing waves can be formed (owing to reflections on the interfaces of tissues with different electric parameters) in cases where the thickness of the considered layer of tissues is comparable with the wavelength (which in turn depends on the electric parameters of the tissue). Table 5, which gives the values of the

TABLE 4. Depth of Penetration of Electromagnetic Waves
into Various Tissues, cm

Tissue	Frequency, MHz							
	100	200	400	1000	3000	10 000	24 000	35 000
Bone marrow	22.9	20.66	18.73	11.9	9.924	0.34	0.145	0.073
Brain	3.56	4.132	2.072	1.933	0.476	0.168	0.075	0.0378
Eye lens	9.42	4.39	4.23	2.915	0.500	0.174	0.0706	0.0378
Vitreous body	2.17	1.69	1.41	1.23	0.535	0.195	0.045	0.0314
Fat	20.45	12.53	8.52	6.42	2.45	1.1	0.342	—
Muscle	3.451	2.32	1.84	1.456	—	0.314	—	—
Whole blood	2.86	2.15	1.787	1.40	0.78	0.148	0.0598	0.0272
Skin	3.765	2.78	2.18	1.638	0.646	0.189	0.0722	—

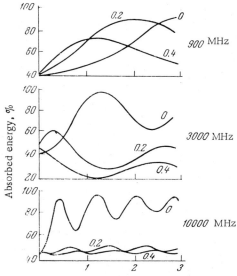

Fig. 17. Absorption of energy of SHF fields of different frequencies in animal tissues in relation to thickness of subcutaneous fatty layer and different skin thicknesses (indicated on curves).

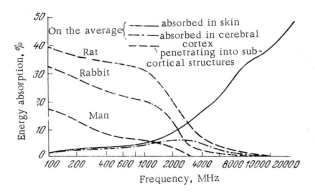

Fig. 18. Frequency dependence of SHF energy absorption in brain structures on irradiation of the head of man, rabbit, and rat.

TABLE 5. Wavelength (cm) in Tissues at Various Frequencies

Tissue	Frequency, MHz							
	100	200	400	1000	3000	10 000	24 000	35 000
Bone marrow	116.1	62.2	32.19	12.63	3.97	1.250	0.368	0.388
Brain	31.7	19.4	11.16	4.97	1.74	0.595	0.200	0.201
Eye lens	33.15	22.3	12.53	5.28	1.75	0.575	0.200	0.201
Vitreous body	21.7	13.0	7.96	3.41	1.18	0.395	0.146	0.154
Fat	96.0	57.1	30.9	12.42	3.79	1.450	0.680	—
Muscle	27.65	16.3	9.41	4.09	—	0.616	—	—
Whole blood	25.15	15.35	8.89	3.87	1.36	0.449	0.214	0.167
Skin	28.07	17.94	10.12	4.41	1.49	0.506	0.250	—

wavelengths in various tissues (Fleming et al., 1961), shows that standing waves can be formed in layers of tissue of man and large animals in EmFs with frequencies above 3 GHz.

4.4. Thermal Effect of EmFs in Tissues of Living Organisms

The heating of the tissues of an animal's body and the general increase in the body temperature due to an EmF depend not only on the amount of electromagnetic energy converted to heat, but also to a great extent on the heat regulation of the organism.

In homeothermic animals (birds and mammals) the resultant heat loss at a given body temperature is equal to the algebraic sum of the heat production due to metabolic processes and the heat loss due to radiation and to evaporation in respiration (and also in sweating in man), as shown in Fig. 19A. In the temperature range within which the organism is still capable of heat regulation — between the points of intersection of the resultant curve and the x axis — heat loss predominates and this leads to restoration of the normal body temperature.

With further increase in temperature the heat transfer may become positive and the body temperature will rise to the lethal point.

These relationships have been demonstrated in experiments in which animals were exposed to EmFs of superhigh frequencies

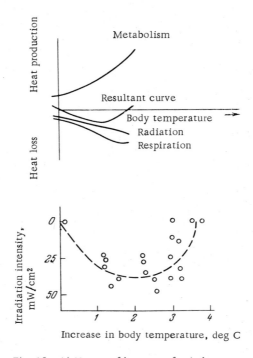

Fig. 19. A) Nature of heat transfer in homeothermic animals in relation to body temperature; B) relationship between intensity of
whole-body exposure of dog to SHF fields and
increase in body temperature.

(Ely and Goldman, 1956a, 1956b) in the temperature range of heat
regulation with automatic maintenance of a prescribed temperature
of the animal (regulation of the irradiation intensity). Figure 19B
shows the resultant curve (similar to the corresponding curve in
Fig. 19A). The maximum possible energy absorption, corresponding to the maximum heat loss, is found when the temperature of
the rat increases by 4.5 deg, that of the rabbit by 3.5 deg, and that
of the dog by 2.5 deg.

Experiments with phantoms simulating an animal's body
(Mermagen, 1961) have shown that with increase in the volume of
the object more time is required to heat it to a prescribed temperature by an EmF of a particular power. This can be attributed,

firstly, to the fact that heating of a large volume requires more calories and, secondly, to the fact that for the same depth of penetration of EmF energy into tissues the fraction of the volume in which absorption takes place is greater, the smaller the volume. For instance, since an EmF with frequency 300 MHz penetrates to a depth of 2.5 cm (for muscle tissues) this means that in the case of a rat (body diameter 5-6 cm) the EmF energy is absorbed by practically the entire body, but in the case of a dog (body diameter 20-25 cm) it is absorbed only in a small superficial part of the body.

A more thorough theoretical investigation of the conditions of heating of the body tissues of man and various animals by microwaves was recently carried out (Hoeft, 1965). The time t required to increase the body temperature by 5 deg (ΔT = 5 deg) was calculated from the equation

$$t = \frac{GC_b\Delta T}{E + M - S_b\alpha_{ab}(\theta_{ab} + \Delta T)}, \tag{35}$$

where G is the mass of the body, C_b is its specific heat, M is the heat due to metabolism, E is the heat due to microwave irradiation, S_b is the surface of the body, α_{ab} is the air−body heat transfer coefficient, and θ_{ab} is the initial temperature difference between the body and the air.

Hoeft concludes that at very low values of t, corresponding to a low irradiation intensity, there is practically no difference in the rate of heating of animals of different size, but at high intensities (t is small) the body of small animals is heated more quickly.

A considerable number of experimental investigations have been devoted to finding out how the heat production in animal tissues is related to the intensity and the time of action of the EmF and to the determination of the temperature distribution in the tissues (Presman et al., 1961). The results of most of these investigations have been contradictory, however. In some cases there was greater heating of the deep tissues than the superficial tissues, in others the temperature distribution was the opposite, and in others there was both a positive and a negative temperature gradient depending on the conditions of exposure to the EmF. The main reasons for these contradictions were the inaccurate determination of the power absorption and the incomparability of several of the experimental conditions.

Reliable results have been obtained in investigations with microwaves in which an improved method* of determining the power absorption was used (Boyle et al., 1950a). Figure 20 shows graphs of the heating of tissues in the thigh of a dog by microwaves. The authors attribute the reduced heating of the subcutaneous fatty layer to low energy absorption and the gradual leveling out of the temperature of the layers to the heat conduction of the tissues and the removal of heat by enhanced blood flow.

Attempts have been made to estimate theoretically the amount of heat released at a prescribed distance from the irradiated surface and to calculate the corresponding increase in temperature (Cook, 1952a, Clark, 1950). However, a comparison of the theoretical and experimental data showed an approximate correspondence only in the case of short exposures to microwaves (Cook, 1950).

An experimental estimate of the threshold EmF intensities required to produce a thermal effect has been carried out in various frequency ranges for the case of whole-body and local exposure of man and animals to EmFs. The threshold of the thermal effect was determined as the minimum increase in the body or tissue temperature which did not exceed the normal fluctuations in the organism. The minimum sensation of the heat has also been used as an index of the thermal effect in man. Incidentally, it has been found (Mittelman, 1961) that the heat sensation and the EmF power absorption in tissues (in the range 20-200 MHz) are connected by the relationship

$$H = \log P - a \log P_0, \tag{36}$$

where H is the heat sensation estimated on a four-point scale (barely perceptible heat, moderate heat, intense heat, almost unbearable heat), P_0 is the absorbed power at which barely perceptible heat is felt, P is the given absorbed power, and a is a constant independent of the frequency (although P_0 varies with the frequency). The results of different estimates are given in Table 6.

As the table shows, the threshold EmF intensities decrease with frequency increase. This is understandable, since the coefficient of absorption of electromagnetic energy is proportional to the frequency and the value of the electric parameters σ and ε, which in turn vary with the frequency.

* This method is described in Section 5.3.

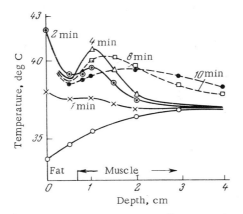

Fig. 20. Heating of tissues at different depths
below surface of dog thigh irradiated by a SHF
field for different periods (indicated on curves).

It should be noted in conclusion that the possibility of selective heating of microparticles in biological media, unaccompanied
by significant heating of the medium surrounding them, has been
discussed several times in studies of the thermal effect of EmFs.
A theoretical analysis (Schwan and Piersol, 1954) has shown, however, that such selective heating is possible only if the particles
are fairly large — not less than 1 mm in diameter. Hence, there
are no grounds for assuming the selective heating of microparticles
(cells, bacteria) in the absence of significant heating of the medium
in which they are suspended.

4.5. Nonthermal Effects of EmFs in

Biological Media

Experimental and theoretical investigations of some interesting microprocesses produced by EmFs have been carried out in
recent years.

The first process of this kind is the formation of chains lying
parallel to the electric lines of force by suspended particles of
charcoal, starch, milk, erythrocytes, and leucocytes on exposure
to continuous and pulsed EmFs of high and ultrahigh frequencies
(1-100 MHz). For each type of particle there is an optimum fre-

Table 6. Threshold Intensities of EmFs for Thermal Effects in Tissues of Living Organisms

EmF range	Organism	Kind of thermal effect	Threshold intensity	Reference
500 kHz	Rats and rabbits	Increase in rectal temperature	8000 V/m 160 A/m	Nikonova, 1964a
14.88 MHz 69.7 MHz	Rats and rabbits	Increase in rectal temperature	2500 V/m 200 V/m	Fukalova, 1964b
Decimeter	Rats and rabbits	Increase in rectal temperature	40 mW/cm^2 (390 V/m)	Lobanova, 1964a
10-centimeter	Man	Sensation of heat	10 mW/cm^2	Vendrik, Vos, 1958
		Heating in irradiated region	10 mW/cm^2 (190 V/m)	Presman, 1957a
	Rats	Increase in rectal temperature	10 mW/cm^2	Lobanova, 1964a
3-centimeter	Man	Sensation of heat	1 mW/cm^2 (61 V/m)	Hendler, Hardy, 1960
	Rats	Increase in rectal temperature	5—10 mW/cm^2 (135-190 V/m)	Gordan and Lobanova, 1960
	Rats	Heating in irradiated region	1.5 mW/cm^2 (75 V/m)	Mirutenko, 1964
Millimeter	Rats	Increase in rectal temperature	7 mW/cm^2 (170 V/m)	Lobanova, 1964a

quency range within which the effect occurs at minimum field strength (Herrick, 1958; Heller, 1959; Heller and Teixeria-Pinto, 1959; Heller and Mickey, 1961; Wildervank et al., 1959).

Theoretical investigations have shown (Satio et al., 1961a; 1961b; Furedi and Valentine, 1962; Furedi and Ohad, 1964) that chain formation is due to the attraction between particles in which dipole charges are induced by the EmF (Fig. 21). In a nonpolar dielectric medium (oil) this effect also occurs at low frequencies and even in an electrostatic field, but in water and physiological saline the ions and dipole molecules shunt the low-frequency field and the effect is possible only at fairly high frequencies (more than tens of MHz). The time constant of chain formation is proportional to the cube of the particles radius (it is 1 sec when the radius is 1μ). It does not depend greatly on E in weak fields and is inversely proportional to E^2 in strong fields. In pulsed EmFs the effect depends on the mean value of E.

Asymmetric particles are oriented either parallel or perpendicular to the lines of force. This depends on the relationship between the conductivity of the particles and the medium surrounding them and on the frequency of the EmF, as Fig. 22 shows (for electric parameters close to biological values). The question of the conditions under which parallel or perpendicular orientation occurs has recently been the subject of a theoretical analysis (Gruzdev, 1965).

The second effect — "dielectric saturation" in solutions of proteins and other biological macromolecules due to intense EmFs of superhigh frequencies — has been considered in Schwan's hypothesis (Schwan, 1958). He suggests that such fields cause all the polarized side chains of the macromolecules to become oriented in the direction of the electric lines of force and that this can lead to breakage of hydrogen bonds and other secondary intra- and inermolecular bonds and to alteration of the hydration zone (on which the solubility of molecules depends). Such effects can cause denaturation or coagulation of molecules, as has been confirmed experimentally (Fleming et al., 1961).

The third effect, due to the action of Lorentz forces in alternating fields on ions in an electrolyte, has been theoretically and experimentally investigated by Heinmets and Herschman (1961). If

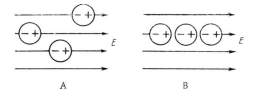

Fig. 21. Orientation of small particles in an EmF.
A) Field applied; B) some time after application
of field.

an electrolyte solution is exposed to electric and magnetic fields which are perpendicular to one another and in phase the electric field (averaged over the time) has no effect on the ions and Lorentz forces cause the positive and negative ions to move in one direction — perpendicular to the direction of the electric lines of force. Effects of this kind have been detected experimentally. In a hemoglobin solution with f = 60 Hz, E = 60 V/cm, and H = 2000 Oe (in amplitude) the colored boundary moved at a speed of 0.36 cm/sec. The authors point out that the considered effects depend on the sum of the ionic mobilities and not on their difference (as, for instance, in the Hall effect) and that such an effect can be produced by an electromagnetic wave propagating in a medium. The action of Lorentz forces in a cellular medium will affect not only the ions of the electrolyte, but also free metabolites in the ionized form.

The resonance absorption of EmFs of various frequency ranges in biological media is of greatest interest.

The absorption and emission spectra of various substances in a wide frequency range — from hundreds of Hz to tens of GHz — are investigated by radiospectroscopy. The investigated features are the energy transitions between closely located energy levels, which include the rotational and inversion levels of molecules, Zeeman splitting of the levels of electrons and atomic nuclei in external and internal magnetic fields, levels formed by the interaction of the quadrupole moments of nuclei with internal magnetic fields, levels of the hyperfine structure of atoms and molecules, and so on (Encyclopedic Dictionary of Physics, 1965). We refer the reader to the special literature on this subject (Gordy et al.,

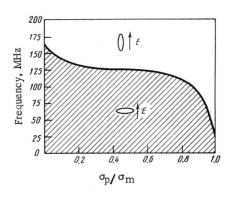

Fig. 22. Orientation of particles with conductivity σ_p, suspended in a medium with conductivity σ_m, in relation to frequency of acting EmF and ratio σ_p/σ_m.

1965; Ingram, 1959, Roberts, 1961; Blyumenfel'd et al., 1962; Free Radicals in Biological Systems, 1963; Setlow and Pollard, 1964) and will merely consider some of the theoretical considerations relating to resonance absorption of EmFs in the tissues of living organisms in vivo and in vitro.

The possibility of resonance absorption of EmFs by protein molecules in connection with the so-called dispersion forces of interaction was recently examined theoretically (Vogelhut, 1960; Prausnitz et al., 1961). The authors proceed from the fact that in proteins, which contain several neutral and negatively charged main side groups, the mean square value of the dipole moment is not zero, even if their mean constant moment is zero. This is due to the fact that (excluding the case of strongly acid solutions) the number of polarized side groups in the protein molecule usually exceeds the number of protons bound with them, so that there is a multitude of possible configurations of the proton distribution in the molecule which do not differ greatly in free energy. In the case of enzyme molecules, on the assumption of a continuous distribution of the main groups, the mean distance between the groups is approximately 9.5 Å. Such dipole–dipole interactions, due to fluctuations in the proton distribution, may be responsible for the absorption of a quantum of energy corresponding to a frequency of 10 GHz. The authors suggest that such a resonance effect of EmFs on the proton distribution in the enzyme molecule can lead to a change in the rate of formation of the enzyme–substrate complex.

The possibility of resonance absorption of EmFs by macromolecules has also been discussed in connection with intramolecular processes. For instance, Barber (1961) suggests that the absorption of the energy of a superhigh-frequency EmF may be due to rotation of the intramolecular structures relative to the C-C bonds, to translational transitions of the hydroxyl groups from one position relative to the hydrogen bond to another, to the rotational levels of metastable states, and so on. The possibility of superhigh-frequency EmFs causing ionization effects which lead to the production of O_2 and OH radicals at high pulsed powers, has also been considered (Tomberg, 1961). These general hypotheses have not yet received convincing experimental verification, but the results of some investigations suggest that such confirmation will be obtained in the near future. For instance, investigations of the prop-

erties of some crystalline proteins, peptides, and amino acids in
the range 1 kHz-4 GHz at temperatures up to 100 deg led Bayley
(1951) to the conclusion that the observed dispersion of the complex dielectric constant is due not only to relaxation of the polar
groups of these macromolecules, but also to resonance absorption
of the EmF.

Resonance absorption has been directly detected in methyl
palmitate on measurement of ε' and ε'' in a wide frequency range —
from 50 Hz to 30 GHz (Jackson, 1949; Dryden and Jackson, 1948;
Cook and Buchanan, 1950). In the region of 4 GHz there was a maximum of tan δ and a plateau on the curve of ε'. The resonance nature of this effect was indicated by the fact that the resonant frequency was independent of the temperature of the solution. The
maximum of tan δ, however, increased with temperature reduction,
which indicated against a purely resonance effect. The author concluded that the observed effect was due not only to resonance absorption of the EmF by the methyl palmitate molecules, but also
to certain relaxation polarization processes.

We will return again to the question of the resonance absorption of EmFs by macromolecules and assemblies of them when we
discuss the mechanisms of the action of EmFs at the molecular
level on the basis of results of recent investigations in this area
(Chapter 10).

The possibility of resonance absorption of EmFs in the whole
body of man and animals or in certain parts of the body has also
been discussed in general. For instance, the loss of control of
motor functions by animals on exposure of the head and spine region to an EmF has been considered from the viewpoint of possible
resonance in the cranial cavity or along the spinal column (Leary,
1959).

In concluding this review of the physical principles of the problem of the biological activity of EmFs we will consider briefly some
of the technical aspects of this problem, viz., the methods of estimating the intensity of the acting fields in medical and experimental
investigations and techniques of protection from the harmful effect
of EmFs of various frequency ranges.

δόσις = a giving, portion, dose

Chapter 5

DOSIMETRY OF ELECTROMAGNETIC FIELDS
FOR THE ASSESSMENT OF THEIR EFFECTS
ON MAN AND ANIMALS

The problem of dosimetry of EmFs in biological and medical investigations involves three main questions:

1. What physical quantities should be chosen for the evaluation of the intensity of the EmF in relation to the conditions and aim of the investigation?

2. What is the relation between the intensity of the EmF measured in the medium surrounding the living object and the "biologically effective" intensity acting directly on the object?

3. What techniques and instruments can be used to measure the intensity of EmFs acting on man and in experiments with animals?

We will consider these questions as they apply to EmFs of various frequencies, beginning with "zero" frequency, i.e., constant electric and magnetic fields.

5.1. Dosimetry of Static Fields

Since $\mu = 1$ in living organisms the intensity of a constant magnetic field measured in the medium surrounding the object can be regarded as the biologically effective value. The intensity H can be measured (in amperes per meter or oersteds) by m a g n e t o - m e t e r s or f l u x m e t e r s (Kalashnikov, 1956; Yanovskii, 1964).

In experimental investigations of animals or objects in vitro they are usually placed either in the gap between the pole pieces of an electromagnet or inside a solenoid, both of which are powered by direct current. In these cases the field strength can be measured and also calculated. The degree of uniformity of the field must be taken into consideration or its gradient at the position of the object must be evaluated. The field can be regarded as uniform in the

central part of a solenoid close to its axis and in the gap between
pole pieces.

Electric field strength in an external medium, determined as
the ratio of the potential difference between two points to the distance between them, is measured by means of e l e c t o m e t e r s with an
electric probe (Kalashnikov, 1956). The estimation of the biologically
effective field strength, however, requires a consideration of the distortion of the field when the object is placed in it. In general this problem
is fairly difficult. The distortion can be assessed approximately
by calculating the dipole moment induced by the field in a conducting body (with the same conductivity as the living object) of a regular geometric shape close to that of the considered object [see formulas (27) and (28)].

In experimental investigations the object is placed between
the plates of a capacitor to which a constant voltage is applied. The
field strength can be measured and calculated (without the object).
The field can be regarded as uniform if the dimensions of the plates
exceed the distance between them.

5.2. Dosimetry of EmFs of Low
to Ultrahigh Frequencies

The intensity of EmFs of low and high frequencies (up to
several tens of MHz) is usually determined in the induction zone in
industrial conditions and in experiments with animals. In this case
the electric field strength E and the magnetic field strength H have
to be measured separately.

The instrument for measuring these values is fundamentally
an ac voltmeter designed for the investigated frequency range and
connected to an appropriate loop antenna for the measurement of
H and a stub or dipole for the measurement of E. The measured
strengths can be determined from the relationships

$$H = a \frac{V}{k_1 S}, \quad E = a \frac{V}{k_2 h}, \tag{37}$$

where V is the measured voltage, a is a coefficient which depends
on the choice of units, k_1 and k_2 are coefficients which depend on the
shape and dimensions of the antenna, S is the cross-sectional area
of the coil, and h is the length of the stub or dipole.

However, the design of such instruments presents some technical difficulties: Firstly, a wide frequency range and a wide range of E and H have to covered; secondly, the action of fields on the voltmeter other than that on the antenna has to be prevented; and thirdly, distortions introduced into the field by the instrument have to be avoided. These difficulties have been overcome to some extent in specially designed Russian-made instruments covering a frequency range from 50 Hz to several tens of MHz and capable of measuring H in the range 0.5–500 A/m and E in the range 5–1500 V/m. Figure 23 shows a general view of one of these instruments — the IEMP-1 (Franke, 1957, 1958).

The distortions introduced by large living objects (the linear dimensions of which are greater than λ) can be estimated approximately with the aid of the formulas (27) and (28) given above.

In experimental investigations of the effects of an alternating electric field the biological object is placed between the plates of a capacitor to which an alternating voltage of the prescribed frequency is applied. The presence of the magnetic component of the EmF must be taken into account in this case. It can be neglected if the dimensions of the object and capacitor are small in comparison with the wavlength λ corresponding to the particular frequency.

In investigations of the effects of an alternating magnetic field the biological object is placed inside a solenoid through which an alternating current of the prescribed frequency is passed. The effect of the electric component can be neglected in the same conditions as in the previous case.

It should be pointed out that the correct setup of experiments in which the action of an alternating electric or magnetic field is investigated separately involves the solution of difficult methodological and technical problems. We will not discuss them here but refer the reader to Solov'ev's investigations (1962, 1963a) as an example.

5.3. Dosimetry of EmFs of

Superhigh Frequencies

The technique of determining the intensity of experimental irradiation in the SHF range involves specific difficulties. In this

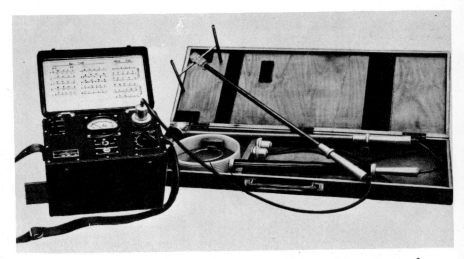

Fig. 23. IEMP-1 instrument for measuring electric and magnetic components of
EmFs in the frequency range 100 kHz to 300 MHz.

region the wavelengths become comparable with the linear dimensions of the electric circuits of the generators and with the linear
dimensions of the biological objects, and in several cases considerably exceed these dimensions. In view of this the components of
SHF genrators and receivers and the methods of exposing biological
objects to these EmFs differ considerably from those used at lower
frequencies (Turlygin, 1952; Presman, 1954a; Valitov and Sretenskii, 1958). We will refer here merely to those special features
of SHF technique which will be encountered later in the description
of the corresponding biological and medical investigations.

1. The electromagnetic vibrations in the SHF range are not
produced in an oscillatory circuit consisting of a capacitor and an
induction coil, but in a closed metal chamber — a cavity resonator.

2. The electromagnetic energy can be transmitted inside
coaxial cables or waveguides (metal tubes or dielectric
rods of circular or rectangular section) with a wave resistance
which depends on the geometric parameters of these transmission
lines.

3. The radiation of electromagnetic energy in wave form can
be directed by the use of horns and reflectors as radiators.

4. In the case of irradiation at a distance some of the energy

is reflected from the object. However, if the object is the load of
the transmission line and its total resistance (impedance) is equal
to the wave resistance of the line, then all the transmitted energy
will be absorbed in it. In general this can be accomplished by including a matching device (impedance transformer) between
the object and the line.

The intensity of the action of SHF fields on human beings
taking part in tests or servicing SHF generators, and also in the
case of long-range irradiation of objects in experiments, is assessed
from the irradiation intensity p, expressed in power flux
density units, usually in mW/cm^2 (Presman, 1956b, 1957b; Schwan
and Li, 1956b; and others).

To compare the conditions of exposure to SHF fields and
lower-frequency fields Knorre (1960) suggests converting the values
of p, E, and H to the value of the bulk energy density of the electromagnetic field (in erg/cm^2) from the known formulas

$$W_E = \frac{10^{-8}}{72 \cdot \pi} \cdot E^2; \quad W_H = 2\pi \cdot 10^{-6} \cdot H^2; \tag{38}$$

$$W_{\text{SHF}} = 3.3 \cdot 10^{-4} \cdot p.$$

The design of instruments for measuring irradiation intensity
in the SHF range is based on the principle of conversion of the energy
of the SHF field (after its calibrated attenuation) to direct current
(by means of a thermistor or crystal detector), the strength of which
gives a measure of the intensity.

Several instruments of such type have been produced in the
USSR and abroad (Presman, 1954b, 1957b; Osipov, 1965; Hirsch,
1956; Mumford, 1961; Vogelman, 1961; and others). Figure 24
shows examples of a Russian and a foreign instrument designed for
a wide frequency range.

In many experimental investigations involving the irradiation
of animals with radiators of various type (Fig. 25) the relative irradiation intensity has been determined from the total irradiation
power and the distance from the radiator to the object. The relative intensity distribution within the irradiated region has been estimated by measuring the increase in the temperature of the surface tissues. Figure 26A shows diagrams of the intensity distribution for radiators with a hemispherical reflector and a radiator
with an angled reflector (Rae et al., 1949). In addition, an attempt
has been made to estimate the power flux density in the center of

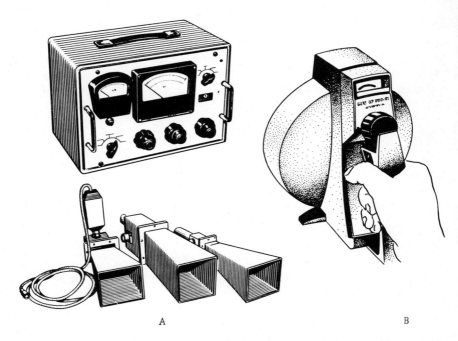

Fig. 24. Instruments for measuring irradiation intensity in SHF range. A) "Medik"
(150-16,700 MHz); B) B86V1 (201-10,000 MHz).

the surface irradiated by a radiator with an angled reflector for
different irradiated powers and distances from the radiator to the
irradiated surface (Fig. 26B) (Hirsch, 1956). Relative dosimetry
of such kind has also been applied with the SHF generators used in
physiotherapy. However, better methods have recently been devised (Livenson, 1962, 1963).

Some methods of irradiating objects in the SHF range such
that all the power is absorbed have been proposed.

One of them is based on the use of a waveguide or coaxial
transmission line which includes a device (impedance transformer)
for matching the total resistance of the object (load) with the wave
resistance of the line and a device to check this matching and measure the SHF power transmitted to the object (Seguin and Castelain,
1947a; Boyle et al., 1950; Presman, 1957c, 1958a; Mermagen, 1961).
Figure 27 shows a diagram of such a setup and the intensity distri-

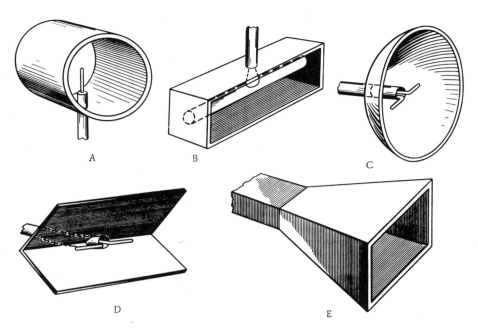

Fig. 25. SHF radiators used in physiotherapy and biological investigations. A) Round waveguide; B) rectangular waveguide; C) with hemispherical reflector; D) with angled reflector; E) rectangular horn.

bution over the cross section of the waveguide or coaxial line. The SHF power transmitted by the line can be absorbed either in the object in vitro, placed inside the waveguide or coaxial line, or in the tissues of part of the human or animal body pressed against the open end of the line. Examples of coaxial and waveguide chambers for irradiation in vitro are shown in Fig. 28.

This method of measuring absorbed SHF power can be used, but is less accurate, in experiments in which one side of the animal's body (dorsal or ventral) is irradiated. In this case an animal of moderate size (rabbit) or a group of small animals (rats, mice) are placed in a special chamber under a horn (or above it) connected to the described transmission line. Figure 29 shows such a device for a rabbit (Presman, 1958a, 1960b).

An example of a method which ensures the absorption of SHF power throughout the body of an animal is that in which the animal

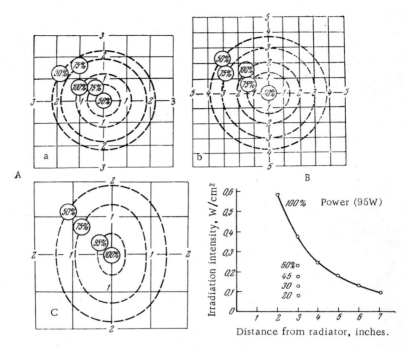

Fig. 26. Distribution of irradiation intensity due to SHF fields delivered by radiators with hemispherical reflectors of diameters 10.2 and 15.2 cm (a and b) and an angled reflector (c); B) irradiation intensity in relation to distance between angled radiator and object.

(rabbit, say) is placed inside the waveguide (at a frequency of 350 MHz) and the difference between the SHF powers entering and leaving the waveguide is measured (Boysen, 1953).

Finally, the possibility of irradiating animals placed in a cavity resonator has been theoretically considered (Eliseev, 1964). This method, however, has not yet found practical application in view of the considerable technical difficulties, although it has been successfully used in experiments with small samples of biological material (Schwan and Windell, 1950; Bayley, 1951).

5.4. Experimental Investigations During the Application of EmFs

The conduction of experimental investigations during the application of EmFs involves considerable methodological difficulties,

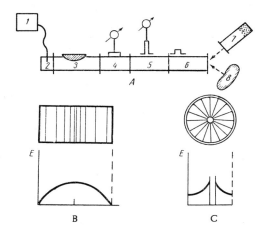

Fig. 27. Diagram of setup to secure absorption in object of SHF power transmitted by a waveguide or coaxial line; B and C) diagrams of distribution of electric field in waveguide and coaxial line. 1) Generator; 2) connection of cable to waveguide; 3) attenuator absorbing a certain fraction of power; 4) meter for measuring power transmitted by line; 5) measuring line for regulating the reflection of the SHF energy from the object; 6) matching device to compensate the reflection of energy; 7) waveguide chamber containing object in vitro; 8) living object.

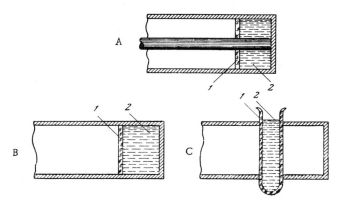

Fig. 28. Coaxial (A) and waveguide (B and C) chambers for application of SHF power to an object in vitro by means of the setup illustrated in Fig. 27. 1) Polystyrene; 2) object.

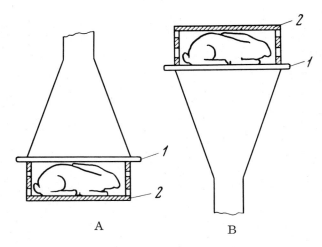

Fig. 29. Device for SHF irradiation of rabbit on dorsal (A) or ventral (B) side. 1) Ploystyrene foam plate; 2) chamber for animal.

especially in the microwave (SHF) range. We will consider only some of the physiological research techniques employed during the application of microwaves.

The difficulty in measuring the body temperature of man and animals during the application of EmFs is due to the fact that mercury thermometers and thermocouples, as well as thermistor gauges, are heated by the direct absorption of EmF energy in them. In early investigations (Esay et al., 1963) benzene thermometers were used for temperature measurement during UHF irradiation, since benzene hardly absorbs any energy from such fields. Recently, however, researchers have begun to use thermistors (Ely and Goldham, 1956) and thermocouples (Boyle et al., 1950; Presman, 1957c) with the wire leads perpendicular to the electric lines of force of the acting EmF. Figure 30 shows a general view of the apparatus which we have designed for measuring the surface temperature on part of the human body during exposure to microwaves by the previously described "contact method" employing a waveguide line (see Fig. 27A). The thermocouple is mounted in the center of the opening of the terminal horn, which is applied firmly to the region of the body. The thermocouple leads are perpendicular to the electric lines of force in the cross section of the horn (see Fig. 27B).

Fig. 30. Device for measuring the temperature of surface tissues on a region of the human body exposed to a SHF field. 1) Measuring line; 2) matching device; 3) rectangular horn radiator; 4) thermocouple galvanometer.

Experiments have shown that in this case the microwaves have hardly any direct effect on the thermocouple.

We have designed (Presman, 1962a) special experimental apparatuses for certain physiological investigations on animals and isolated organs, and for measurements in vitro, during irradiation with microwaves.

The apparatus shown in Fig. 31 can trace the electrocardiogram (ECG) and determine the threshold of electric excitation of the neuromuscular system of a rabbit when its back is irradiated with microwaves (in the frequency range 2.5-3.75 GHz). A measured amount of radiation is delivered by the method described above (see Fig. 29). The ECG is recorded from cuff electrodes on the paws of the animal, which lies freely in the chamber. Screening of the wires leading to the electrodes and the illustrated arrangement of the animal chamber, the microwave generator, and the electrocardiograph ensure the complete elimination of interference from the ECG during the irradiation. The muscular contractions produced by stimulation of a motor point (by rectangular current pulses) are displayed by a specially devised method (Presman, 1963a). Cemented to the animal in the region of the motor point is a miniature laryngophone, which is connected in series with a supply battery and the primary coil of a step-up transformer, the secondary coil of which is connected (through a band filter) to the oscilloscope. This method allows determination of the threshold of electric excitation to within ±15% (with allowance for physiological variations).

Figure 32 shows an apparatus for investigating the electric excitability of a frog nerve-muscle preparation (n. ischiadicus and m. gastrocnemius) during irradiation with microwaves (Presman and Kamenskii, 1961). The preparation, contained in a polystyrene chamber, is mounted in the horn in such a way that only the part of the nerve above the slit between the absorbing plates is irradiated. The stimulating and tapping electrodes are shielded from irradiation by the absorbing plates. This ensures recording of the biocurrents from nerve (or muscle) during microwave irradiation without any interference.

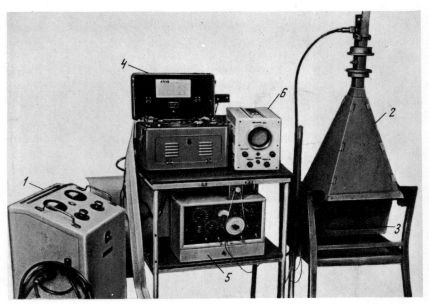

Fig. 31. Apparatus for recording of ECG and determining the threshold of excitation of the neuromuscular system of a rabbit during exposure to a SHF field. 1) SHF generator; 2) horn radiator; 3) chamber containing animal; 4) electrocardiograph; 5) generator of stimulating pulses; 6) oscilloscope display of muscular contractions.

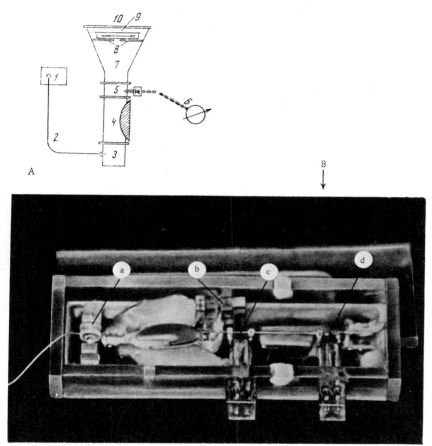

Fig. 32. A) Apparatus for measuring the excitation threshold of a frog nerve-muscle preparation during exposure to a SHF field; B) chamber for nerve-muscle preparation. 1) SHF generator; 2) cable; 3) connector linking cable to waveguide; 4) attenuator; 5) power meter; 6) galvanometer; 7) horn radiator; 8 and 10) absorbing plates; 9) chamber. a) Block for attachment of muscle; b) clamps for holding section of femur; c) tapping electrodes; d) stimulating electrodes.

The discovery of a specific motor reaction of paramecia to electromagnetic stimuli of various kinds (Presman, 1963b, 1963c) gave rise to the need to find a method of applying microwave pulses to paramecia, as well as ac and dc pulses superimposed on a microwave field. Figure 33 shows the apparatus designed for this purpose (Presman and Rappeport, 1965). The paramecia (in a hay infusion)

Fig. 33. Apparatus for investigating the motor reaction of paramecia to a pulsed SHF field or to dc and ac pulses during exposure to a SHF field. 1)SHF generator; 2) dc-pulse generator; 3) ac-pulse generator; 4) case for shielding from SHF field; 5).coaxial chamber for irradiation of paramecia.

are irradiated in a coaxial chamber, which allows the application of stimulating dc or ac pulses during the irradiation and visual observation of the infusoria by means of a binocular microscope.

The apparatus shown in Fig. 34 is designed for measuring the electrical conductivity of liquid samples (protein solutions, for instance) during irradiation either with microwaves or with infrared rays while the temperature of the sample is automatically maintained at a constant level (Presman, 1961, 1963). The part of the tube inside the waveguide line is irradiated. Irradiation with infrared radiation is effected through the opening of the waveguide and irradiation with microwaves through the opening covered by a metal plate. The temperature of heating with infrared radiation is regulated by the closure and opening of the aperture by shield 10 by means of an electromagnetic device 8 and 9. This device is switched on and off by a contact thermometer 3, the bulb of which is submerged in the solution in the part of the tube outside the waveguide and is protected from irradiation by the shield 11. The temperature of the solution during microwave irradiation is controlled by switching on and off the microwave generator by means of the re-

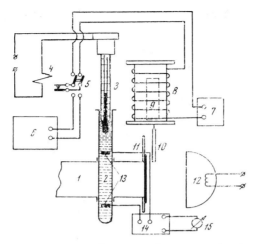

Fig. 34. Diagram of apparatus for comparative investigation of electrical conductivity of liquid biological media during exposure to a SHF field or infrared radiation. 1) Waveguide; 2) test tube containing sample; 3) contact thermometer; 4) normally
closed relay; 5) switch for SHF generator (6) or infrared radiator (12); 7) supply unit for induction
coil; 8,9) iron rod; 10 and 11) infrared shields; 13)
electrodes; 14) instrument for measuring electrical
conductivity; 15) galvanometer.

lay 5 and the contact thermometer. The electrodes for measurement of the electrical conductivity are mounted in the tube outside
the waveguide and, hence, are not exposed to irradiation.

The described pieces of apparatus must be regarded merely
as examples of the approach to the design of methods and equipment for physiological and biophysical investigations during exposure to EmFs. Later, in the discussion of experimental investigations of the biological action of EmFs we will briefly describe the
methods of using this equipment.

5.5. Methods and Means of Shielding

from EmFs

We will briefly discuss the main methods and technical means
of shielding from EmFs in industrial conditions and in the operation
of generators, as well as in the conduction of experimental investigations.

Problems relating to shielding from EmFs have been examined in several monographs (Osipov, 1965; Gordon, 1966) and papers (Gordon and Presman, 1956; Belitskii and Knorre, 1960; Presman, 1958b; Gordon and Eliseev, 1964; Livenson, 1964b; Lang and Koller, 1956; Egan, 1957; Reynolds, 1961; Mumford, 1961; and others). Hence, we will confine ourselves to a discussion of the general principles of shielding and some practical examples.

1. Methods of shielding can be classified as passive or active. The first include such measures as limitation of the time of exposure of persons in the region of the EmF, the location of work points at sufficiently large distances from the EmF source, automation or remote control of operations with EmF sources, and so on. The second include the use of various technical means of reducing the intensity of the EmF to the maximum permissible values laid down in special sanitary directives.* Such technical means can be divided into two groups: shielding devices around the EmF source and individual means of protection (clothing, glasses, etc.) for the personnel operating EmF generators.

2. The intensity of EmFs near their sources can be reduced by:

a) the replacement of components producing EmFs in the surrounding space by special absorbers of electromagnetic energy. For instance, antennas can be replaced by "dummy antennas" of various types;

b) screening of the EmF sources by installing them in a closed chamber made either from sheet metal or wire mesh. In the case of directed emission of EmFs (which is often the case in the SHF range) open screens covered with absorbing material on the emitting side are also used.

The degree of attenuation of the strength of an EmF due to shielding can be calculated from the value of the s c r e e n i n g e f - f i c i e n c y S; for EmFs with frequency up to tens of MHz

$$S_E = \frac{E_0}{E_S}, \quad S_H = \frac{H_0}{H_S}, \tag{39}$$

---

*See, for instance, "Provisional Health Regulations for Work with Centimeter-Wave Generators" in: The Biological Action of SHF, Izd. In-ta gigieny truda i profzabol. AMN SSSR, Moscow (1960), pp. 121-127.

TABLE 7. Values of Coefficient S_1 for Different Frequencies

Kind of screen	Screen material	Frequency, kHz				
		10	100	1000	10,000	100,000
Metal sheets 0.5 mm thick	Steel	$2.5 \cdot 10^6$	$5 \cdot 10^8$	$>10^4$		
	Copper	$5 \cdot 10^6$	10^7	$6 \cdot 10^8$	$>10^{12}$	
	Aluminum	$3 \cdot 10^6$	$4 \cdot 10^6$	10^8	$>10^{12}$	
Metal netting	Copper (wire diameter 0.1 mm, mesh 1×1 mm)	$3.5 \cdot 10^6$	$3 \cdot 10^5$	10^5	$1.5 \cdot 10^4$	$1.5 \cdot 10^3$
	Copper (wire diameter 1 mm, mesh 10×10 mm)	10^6	10^5	$1.5 \cdot 10^4$	$1.5 \cdot 10^3$	$1.5 \cdot 10^2$
	Steel (wire diameter 0.1 mm, mesh 1×1 mm)	$6 \cdot 10^4$	$5 \cdot 10^4$	$1.5 \cdot 10^4$	$4 \cdot 10^3$	$9 \cdot 10^2$
	Steel (wire diameter 1 mm, mesh 10×10 mm)	$2 \cdot 10^5$	$5 \cdot 10^4$	$2 \cdot 10^4$	$1.5 \cdot 10^3$	$1.5 \cdot 10^2$

where the subscripts S and O indicate the strength with and without the screen, respectively.

For the SHF range

$$S_{SHF} = \frac{P_0}{P_S},$$
(40)

where P_0 is the power flux density without the screen, and P_S is the same with the screen.

At frequencies below tens of MHz the value of S for a closed screening chamber can be calculated from the formula (Shapiro, 1955)

$$S = (1.26 \cdot 10^{-7} \cdot Rf) \cdot s_1,$$
(41)

where $R = \sqrt[3]{3v/4\pi}$, v is the volume of the chamber in m^3, and F is the frequency of the EmF in Hz. The expression contained in the brackets represents the dependence of S on the linear dimensions of the chamber and the frequency of the EmF, and S_1 is a coefficient which depends on the screen material. Formula (41) is valid if $R \ll \lambda/6$.

In the SHF range the screening efficiency can be calculated from the formula

$$S_{SHF} = S_1^2, \tag{42}$$

which expresses the attenuation of the power flux density.

Table 7 gives the values of S_1 for EmFs in the frequency range from 10 kHz to 100 MHz for different screen materials, while Table 8 gives the values of S_{SHF} for the 10-centimeter range (2.5-3.75 GHz) for screening by brass wire netting of different dimensions (Presman, 1958b).

We have not touched on the question of electrostatic and magnetostatic fields, since the principles of screening of such fields are known from courses in physics and so far there has been no need to devise special protective devices.

SUMMARY

As we have mentioned in the introduction and as we will attempt to demonstrate in the following chapters, EmFs can have two kinds of effects: They can either regulate the spatial orientation of animals and the rhythm of physiological processes in various organisms, or they can affect the behavior of organisms and their vital processes. From the previously considered (in the first two chapters) physical

TABLE 8. Screening Efficiency (S_{SHF}) of Brass Netting for 10-Centimeter Waves

Number of grid	Wire diameter, mm	Number of meshes per cm^2	S_{SHF}
1	0.53	16	$9 \cdot 10^2$
2	0.43	25	$7 \cdot 10^3$
3	0.35	64	$3 \cdot 10^4$
4	0.25	81	$6 \cdot 10^3$
5	0.20	169	$9 \cdot 10^4$
6	0.14	186	$9 \cdot 10^4$
7	0.075	441	$8 \cdot 10^4$
8	0.08	559	10^5

characteristics of the natural and artificial EmFs encountered in the habitats of organisms we can put forward some *a priori* views regarding which of the considered EmF parameters will be most important in producing these effects.

It is obvious that the effect of EmFs on orientational and navigational systems of organisms must involve directional parameters which remain constant in time or, at least, vary sufficiently slowly. Such parameters are the electric and magnetic lines of force in static fields or in induction EmFs, the vectors E and H in a linearly polarized electromagnetic wave and, finally, the direction of flow of electromagnetic energy in directed emission of electromagnetic waves.

Among the natural EmFs only the earth's electric field is directed perpendicular to the earth's surface. The directional parameters in the magnetic field are the horizontal and vertical components (the dominant component in central latitudes is the horizontal one and in polar regions is the vertical one). It should be borne in mind that the direction of the horizontal component of the magnetic field (declination angle) and also the strengths of both the magnetic and electric field vary with a diurnal periodicity.

We can speak of a definite directivity in artificially produced static fields and in induction EmFs only in relation to a region fairly close to the source. At greater distances from the source there will be the action due to the combination of fields from other sources, the directions of which will generally be different. The same applies to the direction of polarization in electromagnetic waves and their directed emission. In other words, directivity is operative only at a distance from the source where the strength of the EmF created by it is still much greater than the strength of the FmFs at this point due to other sources.

If natural EmFs synchronize the rhythms of biological processes, the main parameter responsible for this regulation must be the period of variation of the EmF.

In natural EmFs we observe two types of periodicity:

1. Periodic changes in strength with a single particular frequency (usually superimposed on a background of some fluctuations); such periodicity is characteristic of the changes in the

earth's electric and magnetic fields and also of the direction of the magnetic lines of force.

2. Periodic changes in the total intensity of noncoherent electromagnetic radiations; such changes are observed in the radiations produced by atmospheric discharges and in cosmic radio emission.

Artificially produced EmFs can be regarded as periodic only at distances from the source where the effect of FmFs produced at this point by other sources and noncoherent with the EmF of the particular source is small. As regards the total intensity of EmFs from different sources generating at different frequencies with a randomly varying phase difference, it is again a case of noncoherent radiations.

If we assume that periodically varying natural EmFs have a synchronizing effect on living organisms, then irregular variations of natural EmFs (during solar flares, for instance) must be regarded as "interference," which will upset to some extent the regulation systems of biological processes. As regards artificially produced FmFs (the frequencies of which in general are different from those of natural EmFs), these fields will either introduce "harmful" information into the biological regulation system (if their parameters do not correspond to the biological rhythms) or can be regarded as "noise," which will adversely affect regulation functions to some extent.

The data given in Chaps. 3 and 4, which illustrate the electric properties of living tissues and the physical principles of interaction of EmFs with biological media, are quite adequate for an examination of the physical mechanisms of the biological effects produced by static fields and FmFs of various frequencies and sufficiently high intensities (magnetomechanic, magnetoelectric, and orientational effects, thermal effects). But are these data sufficient to reveal the mechanisms of the experimentally discovered sensitivity of living organisms to the fields of much lower intensity to which we referred in the introduction?

Looking ahead, we cite some experimental data on the minimum EmF intensities which can produce reactions in living organisms at different levels of their biological organization. As the

Table 9. Comparison of Theoretical Estimates and Experimental
Data on Maximum Sensitivity of Biosystems

EmF range	Theoretically estimated minimum EmF intensity for reaction of biosystem	Experimentally determined minimum EmF intensity for reaction of biosystem
Electrostatic field	10^6 V/m —effect on stability of configuration of macromolecules (Hill, 1958)	10^{-5} V/m —electroreception in fish (Bullock, 1959; Dijgraaf, 1964); $2 \cdot 10^3$ V/m —inhibition of reproduction of Chlorella (Rodicheva et al., 1965)
Magnetostatic field	10^3 - 10^5 Oe. Magnetomechanical and orientational effects (Dorfman, 1966)	0.6 Oe —Increase in motor activity of birds (Él'darov and Kholodov, 1964); 10 Oe — effect on development of cereal plants (Novitskii et al., 1965)
High-frequency EmFs	10^4 V/m —thermal effect in human body tissues (Presman, 1957b)	10^4 V/m —vascular conditioned reflex in man (Plekhanov, 1965): $3 \cdot 10^2$ V/m —change in salivation in dogs (Salei,1964)
Ultrahigh- and super-frequency EmFs	10 mW/cm^2—thermal effect in man and animals (Schwan and Li, 1956a)	$2 \cdot 10^{-2}$ mW/cm^2 —change in EEG in rabbit (Gvozdikova et al., 1964); 1 mW/cm^2 —change in human gamma-globulin activity in vitro (Bach, 1961a)

data given in Table 9 show, the sensitivity to static fields and to
EmFs of various frequency ranges is several orders higher than
the theoretically predicted value. Such high sensitivity is observed
not only in complex organisms, but also in primitive organisms,
in plants, and even at the molecular level (in entire organisms).

Thus, an examination of the mechanisms of energetic interaction of EmFs with biological media cannot provide a basis for
an explanation of the high sensitivity of living organisms to EmFs.

This means that we cannot base our approach to the problem of the biological effects of EmFs solely on the physical principles of interaction of EmFs with the substance, as can be done for nonliving objects. Living organisms apparently have systems which are particularly sensitive to EmFs. These systems can be discovered and their mechanisms revealed only by biological investigations in which due regard is paid to the biological, and not the physical, principles of interaction of EmFs with living organisms. We cannot help recalling Szent-Györgyi's words (1960): "Living Nature also often works with more complex systems than the physicist uses for testing his theories."

It is biological research, employing relatively simple experimental techniques (the main features of which were described in Chap. 5) which has led to the detection of various manifestations of the biological action of EmFs. The second part of this book describes the experimental investigations of the biological action of artificially produced EmFs, while the third part describes the investigation of the regulating and disturbing effect of natural EmFs on life.

Part Two

EXPERIMENTAL INVESTIGATIONS
OF THE BIOLOGICAL ACTION
OF ELECTROMAGNETIC FIELDS

This part deals mainly with investigations of the biological effects
of EmFs carried out in the last 10-15 years. These investigations
cover the entire considered region of EmFs — from constant fields
to millimeter radio waves. The most significant information, however, has been collected in investigations in the UHF and SHF
ranges. Constant magnetic and electric fields and the low-frequency range have been less well investigated and, finally, there has
been relatively little research on high-frequency EmFs. In recent
years there have appeared several collections and review articles
devoted to the biological action of electromagnetic fields in the UHF
and SHF ranges (The Biological Action of Ultrasound and Superhigh-Frequency Electromagnetic Vibrations, 1964; Biological Ef-
ects of Microwave Radiation, 1964; Schwan and Piersol, 1955;
Duhamel, 1958, 1959; Kalant, 1959; Presman et al., 1961; Presman,
1964a, 1964b, 1965a; Gordon et al., 1963; and others), the biological
action of magnetic fields (Questions of Hematology, Radiobiology,
and the Biological Action of Magnetic Fields, 1965; Biological Effects of Magnetic Fields, 1964; Barnothy, 1963a; Becker, 1963a;
Mogendovich, 1965; and others), and the biological action of low-
frequency EmFs (Some Questions of Physiology and Biophysics,
1964).

In the following chapters we will discuss the biological action
of EmFs of various frequency ranges on various levels of biological organization — from the entire organism to cells and molecules
in vitro.

The discussion of the diverse manifestations of the biological action of EmFs in a systematic manner necessitates some form
of classification of these effects. We have tried to do this on the

basis of two main features: Firstly, separate consideration of the
effects of EmFs on entire organisms and in experiments on isolated
cells and solutions of macromolecules in vitro and; secondly, division of the effects of EmFs on the entire organism according to
their general nature into irreversible changes in the organism, organs, and tissues, and reversible changes, manifested in some
fairly rapidly reversible disturbance of physiological functions.
The latter division, of course, is purely arbitrary, since it is
rather difficult to draw a line between reversible and irreversible changes.

Chapter 6

IRREVERSIBLE AND PERMANENT EFFECTS
OF ELECTROMAGNETIC FIELDS IN ENTIRE
ORGANISMS

The lethal effect of an EmF — the death of the living organism
under the action of a high-intensity EmF — is certainly irreversible. The evaluation of any biologically active factor usually be-
gins with an investigation of such experimental effects.

6.1.   Lethal Action of EmFs

Table 10 gives the experimentally determined values of the
intensity and duration of action of EmFs of various frequency
ranges which are lethal (other environmental factors being normal)
for animals of various species.

As the table shows, the death of the animals occurs in cases
where the high-intensity EmF raises the body temperature of the
animal (determined from the rectal temperature) to a level above
the critical value, i.e., to 41-42°C for large animals (dogs and
rabbits), and 42-43°C for small animals (rats and mice).* In
such conditions the thermoregulation of the organism is irrevers-
ibly upset and the animal dies.

---

*Here and henceforth we refer to the white mice and rats usually used in laboratory
investigations.

TABLE 10. Death of Animals Due to EmFs of Various Frequency Ranges

EmF parameters	Animal	Lethal values			Evaluation of lethal outcome		Reference
		EmF intensity	Duration of exposure, min		Temperature increase, °C	%mortality	
Constant magnetic field	Drosophila	120,000 Oe	60		—	100	Beischer, 1964
50-500 Hz		650,000 V/m	60-120		—	70-90	
50 Hz	Mouse	650,000"	270		—	50	Solov'ev, 1963b
500 Hz		650,000"	90		—	50	
14.88 MHz	Rat	9,000 "	10		—	100	
	"	5,000 "	100		—	80	
	"	4,000 "	100		—	25	Fukalova, 1964b
69.7 MHz	"	5,000 "	5		—	100	
	"	2,000 "	100		—	83	
200 MHz	Dog	330mW/cm	15		5	50	
	"	220 "	21		4.1	25	Addington et al., 1961
	Guinea pig	590 "	20		5.9	100	
	"	410 "	20		4.2	67	
	"	330 "	20		4.1	100	Michaelson et al., 1961a
	Rabbit	165 "	30		6-7	100	
2800-3000 MHz (pulsed)	Dog	165 "	270		4-6	100	Michaelson et al., 1961b
	Rabbit	300 "	25		6-7.5	100	Tyagin, 1958
	"	100 "	103		4-5	100	
	Rat	300 "	15		8-10	100	
	"	100 "	25		6-7	100	
	"	40 "	90		—	100	Lobanova, 1960
10,000 MHz (pulsed)	"	400 "	13-14		7	100	Mirutenko, 1964
	Mouse	32 "	1.7		5.6-7.8	100	
	"	8.6 "	33		9.2	100	Baranski
	"	5 "	188		6.8	50	
24000 MHz (pulsed)	Rat	300 "	20		5.5	100	
	"	28 "	139		—	100	Deichman et al., 1959
	Mouse	170 "	6.3		—	100	
	"	37 "	282		—	100	

It is apparent that the tabled values for the increase in body temperature of the animals due to EmFs of the given intensities and durations of action correspond to the theoretical values calculated from formula (32) if the appropriate values of the heat transfer parameters are substituted in it. In other words, the increase in body temperature leading to death of the animal is due to the thermal effect of the EmF with due allowance for the heat transfer of the organism.

However, the course of the processes leading to the death of animals depends on the conditions of exposure to the EmF (the region of the body irradiated, the position of the animal in the field, rate of heating, and so on).

A series of experiments conducted on dogs, rabbits, and rats to investigate the temperature and other physiological reactions to pulsed and continuous SHF fields of high intensity (2800 and 200 MHz, respectively) revealed the following features (Michaelson et al., 1961a, 1961b; Howland et al., 1961).

1. During irradiation of the animals there are three phases of variation of the rectal temperature (Fig. 35) — an initial increase (A), an equilibrium state (B) and, finally, a rapid increase to a value above the critical value, (C), leading to the irreversible failure of thermoregulation and to the animal's death if irradiation is not stopped.

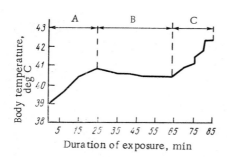

Fig. 35. Increase in body temperature of dog due to whole-body exposure to a high-intensity SHF field (2800 MHz, 165 mW/cm^2). A) Initial heating, leading to increased respiration rate; B) equilibrium state; C) failure of thermoregulation and collapse.

2. With increase in intensity waves with a frequency of 2800 MHz cause a greater thermal effect than waves with a frequency of 200 MHz.

3. Anesthetization of dogs increases the sensitivity to EmFs — at a frequency of 200 MHz the equilibrium phase is shortened and at a frequency of 2800 MHz it disappears. The increase in body temperature due to transfer of an animal to a room with a high temperature (49°C) is not affected by anesthetization.

In rabbits and rats, on the other hand, anesthetization reduces the thermal effect in the case of exposure to a frequency of 2800 MHz, but increases it in rats in the case of 200 MHz.

4. Irradiation of the head of an anesthetized dog (2800 MHz, 130 mW/cm^2) leads to the same increase in temperature as exposure of the whole body to a higher intensity (165 mW/cm^2) for the same time. When a region of the trunk equal in area to the irradiated part of the head is irradiated the thermal effect is much weaker.

5. Long exposure (3-6 h) to fields of intensity 100 and 165 mW/cm^2 leads to leucocytosis (mainly due to an increase in the number of neutrophils), which persists for 24 h after the termination of irradiation, when the rectal temperature becomes normal. There is also a reduction in the number of lymphocytes and eosinophils, usually 24 and 48 h after the end of irradiation. The number of erythrocytes varies in two stages: It decreases in the first 30-60 min of irradiation and then increases and returns to normal 24 h after the end of irradiation.

These results provide very convincing evidence that the death of animals due to EmFs cannot be regarded simply as a result of overheating of the body, since there are several pronounced disturbances in the regulatory processes in the organism, which depend not only on the amount of electromagnetic energy converted to heat, but also on the frequency of the EmF, the region of the body irradiated, and the physiological state of the animal.

The authors came to the conclusion that the described effects can be attributed to the thermal stressor action* of EmFs. In fact, the phases of temperature variation correspond to the three stages of stress — "the alarm reaction," the "stage of resistance," and the "stage of exhaustion," while the observed changes in the blood are characteristic of the early manifestations of thermal stress. However, EMFs of the considered frequency ranges cannot act directly on the peripheral or central nervous systems. Which of these stress stimuli plays the dominant role in the lethal action of EmFs?

---

*Stress, or the general adaptation syndrome, is the set of adaptive reactions of the organism to any harmful factor (heat, chemicals, psychic trauma). The acting agent (stressor) stimulates the activity of the hypothalamus and the release of the hormone ACTH by the hypophysis; this, in turn, stimulates the hormonal activity of the adrenal cortex.

Thermal peripheral stimulation is indicated by the results of investigations carried out at Tulane University (McAfee, 1961; McAfee et al., 1961), where the reactions of cats to local heating of the peripheral nerves (radial, ulnar, trigeminal and sciatic) by a metal heater, infrared radiation, and 3- and 12-centimeter waves (10,000 and 3000 MHz) were compared. The animals were subjected to various kinds of anesthesia, including deep anesthesia. In every case, irrespective of the kind of treatment, an increase in the temperature of the nerve to 46°C led to reactions typical of pain stimulation: accelerated respiration, excitation (arousal from anesthetic state), accelerated heart rate, contraction of the peripheral blood vessels, and increase in blood pressure. Similar reactions were observed when the peripheral nerve was crushed.

Cook (1952c) compared the threshold microwave (3000 MHz) and infrared intensities required to produce an almost intolerable sensation of pain on part of the human body. Such a sensation was produced when the skin temperature rose to 46–47°C. When the irradiated area was fairly large the tolerable microwave intensity (800 mW/cm^2) was four times greater than the corresponding infrared intensity. The energy absorption in the surface tissues (to a depth of 1.5 mm), however, was approximately the same in each case, which can be attributed to the different degrees of absorption. Hence, the effect of local microwave irradiation was apparently due entirely to conversion of the electromagnetic energy to heat.

This would appear to indicate a nonspecific, purely thermal stressor action of EmFs, similar to that caused by heating by infrared rays, for instance. Other comparative investigations, however, show that this inference is by no means clear-cut.

Deichman et al. (1959) discovered different relationships in comparative experiments in which they exposed the back of rats to infrared rays and microwaves of frequency 24,000 MHz (both kinds of radiation are practically totally absorbed in the skin layer). They found that microwaves heated the skin tissues much more rapidly and to a higher temperature than infrared rays of the same intensity: Microwaves raised the temperature to 45°C in 2 min, whereas infrared rays raised it to 42°C in 12 min. A similar difference was noted in the increase of the general body (rectal) temperature, which rose to 42°C in 14 min and to 39°C in 24 min, respectively. Finally, the exposure to infrared radiation re-

quired to produce death was eight times longer than the lethal exposure to microwaves of the same intensity.

These results clearly show that for the same heat production
in the surface tissues of animals microwaves have a much greater
effect on the organism than infrared rays (judged by the increase
in tissue temperature, increase in general body temperature, and
lethal effect). This means that microwaves owe their biological
action not only to the thermal effect. In addition to the thermal
effect microwaves have some other "electromagnetic" effect on the
organism.

This is confirmed by other investigations (Addington et al.,
1961; Deichman, 1961; Deichman et al., 1961), which showed that
the death of animals due to an EmF (200 MHz) depended not only
on the intensity and the duration of exposure, but also on the position of the animal's body axis relative to the plane of polarization
of the wave, on the nature of the modulation of the EmF, and on the
rate of rise of the temperature. When the animals (dogs and guinea
pigs) were placed parallel to the plane of polarization of the wave
the body temperature rose 4-5 deg in 15-20 min and the animals
died, but when they were placed perpendicular to the plane of polarization exposure to the same intensity led to a similar increase
in temperature in 7-10 min but the animals were not killed.

The duration of lethal exposure of rats to continuous microwaves (3000 MHz) was found to be 4-5 times greater than the duration of lethal exposure to pulsed microwaves (Marha, 1963), although it had been found earlier that pulsed and continuous microwaves of the same intensity heat the tissues of animal's body
equally (Scelsi, 1957), but continuous microwaves are more effective than pulsed microwaves as regards raising the general body
temperature (Howland et al., 1961).

A special feature of the lethal effect of microwaves is revealed also in the adaptation of the organism to microwaves by
multiple exposures. Multiple exposures to high intensities (165
mW/cm^2) reduce the sensitivity of the organism: The animals
tolerate longer and longer exposures and it takes a longer exposure for attainment of the critical temperature (Howland et al.,
1961). However, exposures to low intensities (1 mW/cm^2), on the
other hand, sensitize the animal organism (rats) to the lethal action of radiation of higher intensity (65 mW/cm^2) (Marha, 1963).

Thus, while the death of animals due to microwaves is the result of peripheral stress-stimulation, it is not entirely a thermal process. It is possible that all the peripheral excitable structures react to such exposure (it is known that they all respond to electric stimulation).

The death of animals due to EmFs by direct stress-stimulation of the structures of the central nervous system is theoretically possible at frequencies of not more than 1000-3000 MHz (see Section 4.3). This is confirmed by the experimental investigations of Austin and Horwath (1949, 1954), who observed convulsions and the death of rats when their heads were exposed to microwaves with a frequency of 2800 MHz. This raised the temperature of the brain tissues to 40°C, an effect which occurred irrespective of the rectal temperature of the animals. Even more convincing in this respect are the experiments in which monkeys were exposed to EmFs of frequency 225-339 MHz (Baldwin et al., 1960). Only the animal's head, which was held in a cylindrical resonant cavity excited by an antenna in the top of the cylinder, was irradiated. When the head was held with the chin pointing upwards exposure for 1 min led successively to a display of interest, sleep resembling anesthesia, awakening and, finally, pronounced excitation. A three-minute exposure resulted in convulsions and the death of the animal. When the head was fixed with the chin pointing downward there was only excitation and sleepiness, and when the head was free to move there were no disturbances, but the animal raised its head and stared at the antenna. It was characteristic that even in the case of lethal exposure the body temperature was only slightly altered, but in the brain there were disturbances in the interneuronal connections and the rhythm of the biocurrents was slowed down. Irradiation of the whole of the body excluding the head did not produce any reactions or any significant disturbances. The authors regard the described effects as the result of disturbances of the functions of the interbrain and midbrain. This is consistent with the long-known effect of electric stimulation of these brain structures — the impairment of thermoregulation, leading to an increase in the body temperature and to typical stressor reactions.

Thus, the death of animals as a result of direct action of intense EmFs on the brain structures is due not so much to the thermal effect, as to the irreversible disturbance of regulatory functions. Disturbances of such kind occur when there is an abnormally intense flow of afferent pulses into the central nervous system due to general peripheral stimulation.

6.2. Morphological Changes in Tissues and Organs Due to EmFs

Morphological changes in organs and tissues of animals can be produced by a single exposure to EmFs of high intensities or can be cumulative and result from repeated exposures to low-intensity EmFs.

The location of the most pronounced changes corresponds in general to the depth of penetration of the EmF energy. The lower the EmF frequency and the smaller the animal, the deeper the tissues which are damaged. However, less pronounced changes in the deeply located organs and tissues are observed in cases where the EmF is completely absorbed in the superficial skin tissues (on exposure to millimeter waves, for instance).

The morphological changes due to EmFs can be of very diverse nature — from severe injuries in the case of lethal exposures (burns, tissue necrosis, hemorrhages, degenerative changes in the cells, etc.) to moderate or slight reversible changes due to exposure to low-intensity EmFs (moderate vascular changes, slight dystrophic changes in the organs, and so on).

The most thorough and comprehensive investigations of such kind have been made in the SHF range (Pervushin and Triumfov, 1957; Subbota, 1957a; Tolgskaya et al., 1959; 1960; Tolgskaya and Gordon, 1964; Seguin, 1949a; Seguin and Castelain, 1947a, 1947b; Boysen, 1953; and others). The severe morphological changes found after exposure to intense irradiation are similar to those which occur when the tissues are overheated by ordinary methods (convective heat, infrared rays).

Of particular interest are the morphological changes which result from multiple exposures to EmFs of low intensity, starting at intensities of about 1 mW/cm^2. Nerve tissues — the skin receptors — are most sensitive in the case of exposure to millimeter and 3-centimeter waves, and the interceptors and structures of the central nervous system are most sensitive to 10-centimeter and decimeter waves.

We must consider separately the structural changes in the brain cells due to multiple exposures to low-intensity SHF fields (Tolgskaya et al., 1960; Tolgskaya and Gordon, 1964). These are, firstly, the changes in the interneuronal connections of cells of the cerebral cortex, manifested in the dissappearance of the spinules* and the appearance of beads on them, and sometimes their breakup into fragments. Secondly, there is the enhanced multiplication of the microglial cells and the slight dystrophic changes in the processes of these cells. The changes due to exposure to millimeter waves are of a focal nature; exposure to centimeter waves leads to enhanced multiplication of the cells around the brain vessels; decimeter waves cause dystrophic changes.

Changes of a similar kind are produced in the brain structures by exposure to high-frequency EmFs and even to a constant magnetic field.

Repeated exposure of rats to EmFs of frequency 500 kHz and nonthermal intensities (1800 V/m and 50 A/m) led to changes in the synapses of the cortical cells and in the structures of the thalamo-hypothalamic region and the brain stem (Tolgskaya and Nikonova, 1964), while prolonged exposure of animals (rabbits, cats, rats) to a constant magnetic field of 200–300 Oe led to dystrophic changes in the neuroglia† (Aleksandrovskaya and Kholodov, 1965, 1966). A longer exposure of guinea pigs to a more intense magnetic field (7000 Oe, 500 h) led to considerable morphological changes (hemorrhages, edemas, necrobiotic processes) in several organs — lungs, spleen, gastrointestinal tract, and testicles (Ryzhov and Garganeev, 1965; Toroptsev et al., 1966a).

A summing-up of the results of investigations of multiple exposures to low-intensity EmFs leads to the following conclusions:

---

*The spinules are the pyriform protoplasmic outgrowths on the apical dendrites of pyramidal cells. According to most recent ideas, the spinules are the receptor apparatus of the cerebral cortex.

† The neuroglia is auxiliary brain tissue. It was formerly believed that the glia cells performed only supporting trophic functions. It has recently been established that the glia cells are implicated in the conduction of nerve impulses and are possibly "memory cells." The microgliar cells behave like typical phagocytes, i.e., they apparently perform a protective function in the nervous system.

1. Morphological changes in tissues and organs are produced by EmFs of various frequencies and a constant magnetic field in the absence of any significant thermal effect. They are presumably due to progressive functional deterioration in the regulation of metabolic processes.

2. The most frequently observed morphological changes are observed in the tissues of the peripheral and central nervous system and impair its regulatory functions by disruption of the corresponding links or by alteration of the structure of the nerve cells themselves. The disturbances caused by EmFs of various frequencies and a constant magnetic field are of the same type.

There is also convincing experimental evidence that even the severe morphological changes produced by high-intensity EmFs cannot be attributed entirely to the thermal effect. This is exemplified by the numerous investigations on the effect of intense EmFs on the eyes and testicles of animals.

6.3. Action of EmFs on Eyes and Testicles

The eyes and testicles are organs which are poorly supplied with blood vessels. Hence, they will be heated more strongly by EmFs than the organs in which the heat can rapidly be removed by enhanced blood flow.

Several experimental investigations have shown that cataracts can be produced by SHF fields (Radiation Cataracts, 1959; Belova and Gordon, 1956; Belova, 1960a; Williams et al., 1956; Merola and Kinosita, 1961; Carpenter et al., 1960, 1961; Van Ummersen and Cogan, 1965; and others). The following main features of this effect have been established:

1. A single exposure of the eyes to microwaves (3- to 30-centimeter) of high intensity leads to the development of a cataract (opacity of the lens) immediately after exposure or after a longer period — up to ten days. The threshold intensities for cataract induction depend on the duration of exposure (120 to 700 mW/cm^2 for 270 to 5 min, respectively).

2. The nature of the development of the cataract, its location, and severity are practically the same for all the indicated conditions of irradiation and do not depend greatly on the micro-

wave frequency. It has been observed, however, that pulsed microwaves are more cataractogenic than continuous microwaves.

3. Cataracts can also be produced by repeated exposures and the threshold intensity and duration of exposure are shorter than in the case of a single exposure (for instance, ten 30-min exposures at an intensity of 150 mW/cm^2). Finally, there have been a few cases of cataract produced in man by chronic exposure (for several years) to microwaves with an intensity of a few mW/cm^2.

In early investigations the cause of the cataract was presumed to be thermal coagulation, but it became more and more obvious from subsequent experiments that this effect is due to the disturbance of metabolic processes and to biochemical changes in the lens. For instance, exposure of the eyes of rabbits to microwaves with an intensity below the threshold for cataract induction and simultaneous injection of alloxan (which causes cataract), also in a subthreshold dose, gave results which indicated the disturbance of metabolic processes in the lens. The authors (Richardson et al., 1952) expressed the hypothesis that a change in glutathione content plays a significant role in this effect. This hypothesis was confirmed by direct observations (Gapeev, 1957; Carpenter et al., 1960; Merola and Kinosita, 1961) which showed a reduction in the glutathione and ascorbic acid content of the eye before the development of radiation cataract and even in cases where no cataract was produced. Characteristic features were an initial reduction in ascorbic acid content (within 18 h), followed by a reduction in glutathione content (within 48 h), whereas in the case of cataracts due to other causes (ionizing radiation, alloxan diabetes, etc.) the initial feature was the change in glutathione content.

Microwaves lead to other biochemical changes in the crystalline lens: reduction in the activity of the enzymes adenosine phosphatase and pyrophosphatase (Daily et al., 1951) and the formation of three types of radicals (Duhamel, 1959). Finally, rabbits exposed daily for 3.5 months to microwaves with an intensity of 1 mW/cm^2 showed a reduction in intraocular pressure; the author (Belova, 1960b) regarded this as due to defective operation of the particular control system.

Early investigations showed that degenerative changes in rat testicles following a 10-minute exposure to microwaves (2800 MHz) occurred when the temperature of the tissues rose to 30-35 deg,

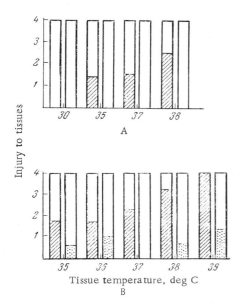

Fig. 36. Injury to rat testicles (evaluated on a four-point scale) due to heating to various temperatures by SHF (left-hand columns) and infrared (right-hand columns) irradiation. A) One h after irradiation; b) 2 days after irradiation.

whereas similar changes were produced by infrared irradiattion only when the temperature rose to 40 deg (Imig et al., 1948). Recent investigations (Searle et al., 1961) have shown an even more significant difference between the effects produced by these two thermogenic factors. The degree of injury to rat testicle tissues is illustrated diagrammatically in Fig. 36, which shows that the effect of microwaves is more pronounced and is manifested earlier than the effect of infrared rays.

Changes in the testicles have also been found after multiple exposures of animals to microwaves. Exposure to 3-centimeter waves with intensity 100 mW/cm^2, which caused an increase of only 3.3 deg in the temperature of the testicular tissues, led to atrophy of the seminiferous tubules (Prausnitz and Susskind, 1962), while exposure to 10-centimeter waves with an intensity of only 0.3 μW/cm (!) for 2-3 months led to changes in the epithelial and interstitial cells (Dolatkowski et al., 1963, 1964).

Exposure of guinea pigs to a constant magnetic field (7000 Oe, 500 h) led to morphological changes in the testicles in the form of necrobiotic changes in the cells of the spermatogenic epithelium; a reduction in their RNA and DNA content was observed (Toroptsev and Garganeev, 1965).

Thus, the nonthermal action of EmFs is indicated by the cumulative effects of multiple exposures to low-intensity EmFs, by the effects due to a single exposure without significant heating of the tissues, and by the effects of a constant magnetic field. It is obvious also that the pronounced morphological changes in eyes and testicles due to high-intensity EmFs cannot be attributed entirely to the thermal effect. We again encounter here some "electromagnetic effect," which plays a significant role in the induction of structural changes in tissues and in the disturbances of metabolic processes.

In this light investigations of the effect of EmFs on tissues and organs of animals with pathological disorders are of great interest. Such investigations have been concerned mainly with malignant tumors and radiation injuries.

6.4. Effects of EmFs on Malignant Tumors and Radiation Injuries

The first investigations of the effect of SHF fields on malignant tumors were carried out as far back as 1938 in Holland by Van Everdingen (1938, 1941, 1946), whose two main lines of research were concerned with the effects of irradiation of animals and the effects of injection of substances which had been irradiated.

Exposure of mice with a cancerous tumor to microwaves of frequencies 460 and 1200 MHz did not produce any results. Irradiation of diseased animals (with lymphosarcoma, Degtyar cancer) with frequencies of 1870 and 3000 MHz, however, was more effective: Daily 5- to 10-min exposures for 14 days inhibited the growth of the tumor, but it began to grow again after the end of the treatment.

The injection of extracts (in carbon bisulfide) of malignant fatty and skin tissues which had been exposed to microwaves with a frequency of 3000 MHz into diseased animals temporarily retarded tumor growth and increased the life span in comparison with the control. These effects were not observed in every case, however, and sometimes the opposite effect was observed. Much more definite results were obtained when irradiated (for 5 min) extracts of tissues taken from healthy animals were injected into diseased animals. Such injections always led to inhibition of tumor growth if fats were excluded from the animals' diet during the period of treatment. Injections of an irradiated glycogen solution, on the other hand, accelerated tumor growth. Van Everdingen believed that these effects were due to molecular changes in the irra-

ΣΟΚΟΣ = leathern bag, pouch, bladder

diated substances. He noted that fat and skin extracts, which are usually optically inactive, became dextrorotatory after irradiation, whereas the optical activity of glycogen, on the other hand, was reduced until it almost disappeared.

Injections of an irradiated tripalmitin emulsion into diseased animals (fed on a fat-free diet) inhibited tumor growth and the malignification of papillomas and led even to the complete disappearance of papillomas.

Another early investigation (Montani, 1944) showed complete resorption of a sarcoma in rats after exposure to microwaves of frequency 6000 MHz and very low intensity.

Inhibition of the growth of malignant tumors in mice due to the action of microwaves (3000 and 10,000 MHz) has also been observed in the case of intense irradiation, leading to considerable heating of the tissues (Roberts and Cook, 1952).

The action of a constant magnetic field on the growth of a cancerous tumor (Ehrlich's ascites cancer) in mice gave negative results — the mortality rate was increased (Gross, 1962). The combined action of a magnetic field and microwaves, however, was very effective. We must discuss more fully these interesting investigations (Riviere et al., 1965a, 1965b, 1965c, 1965d).

Rats with a transplanted uterine tumor (T8e6 atypical epithelioma) and with a transplanted lymphosarcoma (347 and LS-2) were subjected to a simultaneous exposure to a constant magnetic field (300 and 620 G) and microwaves of a particular frequency (3-80 cm) modulated with some frequency in the range 17-300 MHz.

In the experiments with the epithelioma the treatment began on the 2nd, 6th, 10th, or 14th day after implantation and daily exposures of 10-90 min were given for 25-27 days. In the experiments with the lymphosarcoma the treatment began on the 2nd, 5th, or 7th day and daily exposures of 80-140 min were given for a month. In each series of experiments groups of 12 animals and a control group of 24 were used.

Treatment begun even on the 14th day after transplantation of the epithelioma led to the disappearance of the tumor and metastases. Similar results were produced by irradiation in the case of lymphosarcoma, even when the treatment was begun on the 7th day after transplantation. It should be noted that there were no secondary reactions and no relapses in two to three months after the course of treatment.

Figure 37 illustrates the results of exposure to EmFs in experiments with LS-2 lymphosarcoma.

The described investigations were begun quite recently — in 1965 — but the results obtained can be regarded as very promising. There is no doubt that the further development of research on the combined action of EmFs on malignant tumors will be of great importance for cancer therapy and biology.

Very diverse kinds of investigations of the combined action of EmFs and ionizing radiations have been carried out. The above-mentioned investigations of Michaelson et al. (1958) showed that dogs which had been subjected to ionizing radiation (LD50 and LD80) reacted much more strongly to subsequent exposure to high-intensity microwaves: The critical body temperature was reached more rapidly, the number of erythrocytes in the blood underwent greater variation, and so on. Yet animals exposed to ionizing radiation reacted in the same way as nonirradiated animals to a period in a room at 49 deg. The authors believed that these results indicated some physiological interrelationship between the effects of ionizing radiation and microwaves and they undertook special investigations (Michaelson et al., 1963; Howland and Michaelson, 1964; Thomson et al., 1965), which gave the following results:

1. When dogs were exposed simultaneously to pulsed microwaves (2800 MHz, 100 mW/cm^2) and gamma rays (250 kV, 2 r/min, 0.65 mm Al) the number of leucocytes returned to the normal level much sooner after irradiation than in the absence of microwave treatment. Immediately after the combined irradiation the number of leucocytes was doubled, instead of being reduced by 6% as in the case of exposure to ionizing radiation alone. These effects were most pronounced in the case of neutrophils (Fig. 38). The mortality in the case of combined treatment was the same as in the case of exposure to ionizing radiation alone.

2. Prior exposure to microwaves for 3-4 h reduced the mortality due to ionizing radiation (whole-body exposure, 340 r) by a factor of almost four, whereas a shorter exposure to microwaves

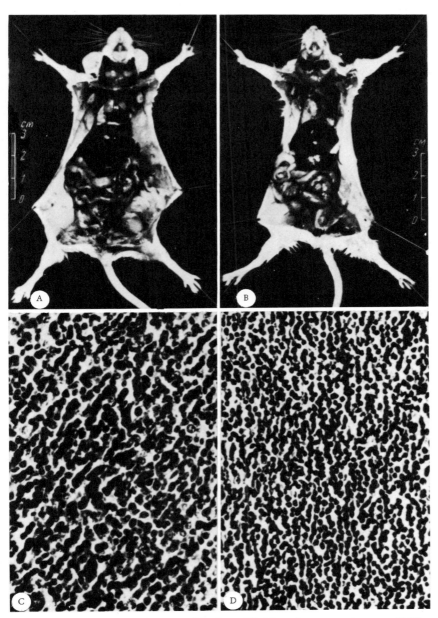

Fig. 37. Effect of combined action of EmFs on LS-2 lymphosarcoma in rats. A) View of lymph nodes before treatment (ten days after transplantation); B) view of nodes in rat after 30 sessions of treatment; C) histological section of inguinal node before treatment (X440); D) the same after treatment.

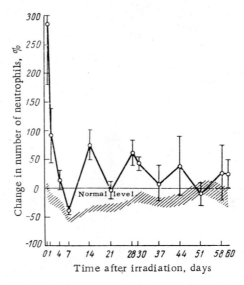

Fig. 38. Change in number of neutrophils in dog's blood after simultaneous exposure to a SHF field and gamma rays in comparison with the corresponding change due to gamma-ray exposure alone (hatched area).

(0.5-1.5 h) had no effect. The percentage mortality was not altered by more intense (165 mW/cm^2) or longer (2.7-5 h) microwave irradiation. Finally, multiple exposures to microwaves of either intensity (for a total period of 4.5-45.5 h), to which the animals became adapted, reduced the mortality due to ionizing radiation by a factor of approximately one-and-a-half.

3. In the case of local exposure of the back region in the tailward direction to ionizing radiation in a dose of 950 r prior exposure to microwaves (100 mW/cm^2, 3-6 h) reduced the mortality from 38% to zero.

Less significant results were obtained when the head of the animals was irradiated: In the case of doses of 5000 and 10,000 r microwave irradiation had no effect on the life span and it was only in the case of a dose of 2500 r that there was some increase in the life span (from 22 to 43 h).

4. The percentage mortality among mice after exposure to microwaves (100 mW/cm^2 for 10 min or for 10 min for a period of 14 days before the death of 30% of the animals) and 14–30 days after exposure to gamma rays was reduced only in the case of the 800-r dose (from 75.9 to 40), and was unaffected at doses of 700 and 900 r.

The authors regarded these effects as indicative of the stressor action of microwaves, which is characterized, as already mentioned, by an increase in the number of leucocytes, especially neutrophils. They emphasize, however, that the increase due to the simultaneous action of microwaves and ionizing radiation was 16 times greater (!) than that due to microwaves alone. The fact that the increase in the number of leucocytes and its rapid return to the normal level occurred only in the case of simultaneous irradiation or when microwave irradiation immediately followed exposure to ionizing radiation indicates that for a short critical time after exposure to ionizing radiation* microwave irradiation can apparently stimulate leucocyte production. The authors also drew attention to the reduction in the partial pressure of oxygen in the arterial blood after exposure to microwaves. This can be regarded as an index of enhanced resistance to ionizing radiation.

A much less pronounced effect of microwaves on the resistance of animals to ionizing radiation has been found in other investigations (Presman and Levitina, 1962c): The mortality of rats which had been exposed to continuous microwaves of low intensity (2800 MHz, 10–15 mW/cm^2) for 30-min periods for 25 days was reduced by two-thirds in comparison with the normal level due to subsequent exposure to ionizing radiation (600 r, gamma rays). Exposure to pulsed microwaves of the same mean intensity with other conditions equal did not alter the percentage mortality in comparison with the control. This effect, of course, cannot be attributed to the thermal stressor action of microwaves, since their intensities were too low. Yet, as we will show later on, the number of leucocytes is increased even by the action of microwaves of nonthermal intensities. This may be due to the direct action of microwaves on the functions of the thalamo–hypothalamic region of the brain, which regulates the number of leucocytes in the blood.

---

*Michaelson et al. (1963) found that the bone marrow continued to produce leucocytes for a short period after exposure to ionizing radiation.

We must mention another effect of simultaneous exposure to ionizing radiation (2620 r, 220 kV) and microwaves of frequency 3000 MHz (Cater et al., 1964). Tumor growth in rats with a transplanted liver tumor following such treatment (the tumor was heated by the microwaves to 47 deg) was much slower and death was much later than after exposure to each of these factors separately.

Investigations with a constant magnetic field have also been carried out. In some experiments (Barnothy, 1963b) mice were subjected to the action of a field of 4200 Oe for 14-20 days and were then exposed to gamma rays (800 r, 33-100 r/min). The mortality in this case was reduced by 26.7 ± 3.2% on the average in comparison with mice which had not been exposed to a magnetic field, and the number of leucocytes was 13% of the normal on the 10th day after gamma irradiation (whereas in mice subjected only to gamma irradiation it was 9% of the normal). Other investigations (Amer and Tobias, 1965) have shown that the percentage of morphological changes in the wing of the flour beetle *Tribolium confusum* due to ionizing radiation is reduced if a magnetic field of 8000 G is applied at the same time. This protective effect depended on the temperature and on the partial pressure of oxygen. The authors believe that a magnetic field affects hormonal regulation.

A slight reduction in the lethal effect of ionizing radiation (900 r) on mice was observed when a low-frequency electric field (50 Hz, 8-12 V/cm) was applied during irradiation (Moos, 1964). In the period between the 5th and 11th days after such treatment the number of deaths was 20-25% less than after exposure to ionizing radiation alone.

Thus, in experiments on the action of EmF on malignant tumors both the nature of the observed effects and their dependence on the EmF parameters indicate that we are dealing here not with the thermal effect of EmFs, but with their effect on the regulatory functions in the organism and the regulation of intracellular processes. This is confirmed by corresponding investigations in vitro, which will be described in Chapter 9. In the case of radiation injuries we find that EmFs affect the regulation of hemopoiesis and other neurohumoral regulatory systems in the organism.

In concluding this chapter we must point out that in most cases the discussed irreversible and permanent effects of EmFs were observed after exposure to high intensities. If in these cases

some nonthermal "electromagnetic" aspect of the action of EmFs
was manifested, it occurred against a background of the thermal
effect, which was often predominant, and sometimes even masked
the nonthermal effects. Yet, as we have seen, nonthermal effects
do occur: They are revealed by investigations with low-intensity
EmFs, where heating of the tissue is insignificant or absent altogether.

Chapter 7

EFFECT OF ELECTROMAGNETIC FIELDS
ON NEUROHUMORAL REGULATION IN
ENTIRE ORGANISMS

The effect of EmFs on neurohumoral regulation is revealed both in
externally manifested reactions and in changes in the nature and
rate of physiological processes. The first types of effect include
changes in the behavior of animals: unconditioned reactions to the
EmFs and changes in previously elaborated conditioned reflexes;
the second type includes changes in the functions of different divisions of the nervous system, disturbances of humoral regulation,
changes in the nature and rate of biochemical processes, and so on.

The effect of EmFs of various frequencies on neurohumoral
regulation has been the subject of numerous investigations for a
long time. This aspect of the biological action of EmFs has been
dealt with in several review papers and monographs (Livshits,
1957a, 1958; Gordon, 1964a; Kholodov, 1965a, 1965b; Frey, 1965).
We will discuss mainly the investigations of recent years.

7.1. Effect of EmFs on Behavior
of Animals

The effect of EmFs on the behavior of animals is revealed
in changes in general motor activity, attempts of the animal to es-
cape from the acting factor, and orientational reactions to EmFs.

The motor activity of mice exposed to a constant magnetic
field of several thousand oersteds was increased by about 50%
(Barnothy and Barnothy, 1960). A much more pronounced effect,

however, was observed in experiments with birds of the sparrow family (Él'darov and Kholodov, 1964; Él'darov, 1965). The birds were placed in a magnetic field of only 0.6-1.7 Oe and their motor activity between 9 a.m. and 7 p.m. was recorded over a period of several days. The experiments (with appropriate controls) were conducted under three kinds of illumination — natural, weak artificial, and strong artificial.

The magnetic field increased the total motor activity by a factor of 3.5-4 in comparison with the control and even by a factor of tens in the evening hours. The distribution of the intensity of this activity in relation to the time of day was also altered: Instead of a gradual reduction towards evening under natural illumination there was a second maximum at 3 p.m. In weak artificial light the activity reached a maximum at noon and in strong artificial light there was one maximum at 2 p.m., instead of two at 11 a.m. and 5 p.m.

A magnetic field also increases the motor activity of fish. In experiments with sticklebacks the activity was increased by 64% by a field with a strength of 50-150 Oe (Kholodov, 1958).

Direct motor reactions to an electric field have been observed in mice (Solov'ev, 1963b): A pulsed field with a strength of 100 kV/m led to synchronous motor reactions. Moos (1964) observed the effect of an electric field on the motor activity of mice: The activity of animals exposed to a field of 8-12 V/cm was increased by 69% during the night, and was increased by 49%, or reduced by 36%, during the day. An increase in the motor activity of rats exposed to EmFs of 450-950 MHz has been reported (Eakin and Thomson, 1962).

The observations of the motor activity of insects subjected to the action of an electric field are of great interest (Edwards, 1960a). The activity of fruit flies (Drosophila) was temporarily reduced by sudden exposure to a field of 10-62.5 V/cm. The period of reduced activity was prolonged if the polarity of the field was reversed at 5-min intervals. A briefer reduction of activity at higher field strengths was observed in bluebottles (Calliphora). Edwards draws attention to the known fact that dipterous insects become more active just before thunderstorms. However, he attributes this effect not to the increasing field strength, but to the increase in currents in the atmosphere, i.e., to the increase in the number of aeroions. This was confirmed by the results of corresponding experiments (Edwards, 1960b) in which the action of an ion current of density $5.2 \cdot 10^{-14}$ A/cm^2 led to enhanced motor activity of bluebottles in approximately 40 min. In addition, other authors report that a change of ion currents in the atmosphere from $3 \cdot 10^{-16}$ to $3 \cdot 10^{-11}$ A/cm^2 leads to increased motor activity of these insects.

Avoidance reactions to a low-frequency EmF (1-10 kHz, 9 V/cm) have been observed in the golden hamster (Schua, 1953), and dogs avoid exposure to SHF radiation (2800 MHz, 40-165 mW/cm^2) (Michaelson et al., 1958). Orientation in the direction of SHF radiation with frequencies of 2800 and 24,000 MHz has also been observed: Dogs turned their heads towards the radiation source and rats oriented their bodies in this direction.

Interesting orientational reactions have been observed in ants (Jaski, 1960). In the zone of action of a SHF field (10,000 MHz) the insects oriented their antennas along the electric lines of force and lost the ability to "communicate" the location of food to other ants. It is of interest that the length of the antennas of the large ants used in these experiments was almost a quarter of the wavelength.

We will mention here the series of investigations (Brown, 1962b) which revealed the ability of planarians and snails to orient themselves relative to the earth's magnetic field (these investigations are fully described in the third part of the book). Finally, numerous experiments have shown motor and orientational reactions of unicellular organisms to an electric field and EmFs of a wider range of frequencies — from low to superhigh frequency (these effects are described in Chapter 9).

The discussed motor reactions of animals must obviously be attributed to the action of EmFs on the nervous system, since motor activity, the defense reaction (avoidance), and the orientation of animals are usually produced by the action of various factors either on the sense organs or directly on different divisions of the nervous system. However, more convincing evidence of the action of EmFs on the nervous system has been provided by investigations of conditioned-reflex and unconditioned-reflex reactions.

7.2. Effect of EmFs on Conditioned

and Unconditioned Reflexes

The effect of EmFs on conditioned reflexes has been investigated in two main ways: 1) EmFs have been used as the conditioned stimulus for the elaboration of conditioned reflexes; 2) the effect of EmFs on conditioned reflexes produced by other conditioned stimuli has been investigated.

Experimental attempts to use a magnetic field as a conditioned stimulus to elaborate a conditioned reflex in amphibians (frogs), birds (pigeons and bullfinches), and mammals (rabbits) were unsuccessful. However, a magnetic field acted as a conditioned stimulus on fish (Kholodov, 1958): In response to a brief exposure (10-20 sec) to a field of 10-30 Oe fish (carp, sticklebacks, etc.) built up a conditioned defense reflex. It was suggested earlier (Lissman, 1958) that the lateral line of fish is the organ which is sensitive to a magnetic field. Further investigations, however, (Kholodov, 1963a) did not confirm this hypothesis. It was found that denervation of this organ did not affect the elaboration of the conditioned reflex. The effect was also unaffected by injury to various brain structures (forebrain, cerebellum, visual cortex). Only injury to the interbrain affected the reaction of the fish to a magnetic field.

Conditioned reflexes can be elaborated in fish to very small changes in electric field gradient (Lissman and Machin, 1958) and to electric pulses of varying length (from hundredths of a millisecond to several milliseconds) and repetition rate (from one per sec to hundreds per sec) (Lissman and Machin, 1963).

A conditioned defense reflex to the application of a low-frequency (200 Hz) EmF produced by a vibrator mounted above the subject's head can be elaborated in man (Petrov, 1952). This reflex, however, is unstable.

Of great interest are the investigations in which a conditioned vascular reflex (detected by plethysmography) has been elaborated to a high-frequency (735 kHz) field of very low intensity — from $2.2 \cdot 10^{-4}$ to $3.3 \cdot 10^{-4}$ V/m (Plekhanov, 1965; Plekhanov and Vedyushkina, 1966). The elaboration of the reflex required 13-25 combinations of the EmF with the unconditioned stimulus (cold) and the reflex persisted for a fairly long time.

An effect of a low-frequency EmF on the behavior of mice in a T-shaped maze has been detected (Amineev and Sitkin, 1965). The animals entered the maze by passing along the axis of a solenoid supplied by an alternating voltage of 100 Hz, which produced a field of 300-470 Oe on the axis. This treatment did not affect the learning to turn in a particular direction, but the reteaching of the animals was 36-40% more rapid in the presence of the field than in its absence. The authors regarded this as a manifestation of the disinhibiting effect of EmFs.

A SHF field can be used as a conditioned stimulus (Makhalov et al., 1963). A conditioned reflex to a brief (20-sec) exposure to a SHF field with a frequency of 2200 MHz and intensity 20 mW/cm^2 was elaborated in mice after 30-50 combinations with the unconditioned stimulus (electric stimulus). The reflex was extinguished by four to seven treatments without reinforcement.

Much more experimental data have been obtained in investigations of the effect of EmFs on previously elaborated conditioned reflexes. The nature of this effect depends on the kind of animal, the type of higher nervous activity, the parameters of the acting EmF, and the exposure conditions.

Some researchers have carried out a series of experiments in which rats were subjected to a single exposure of pulsed SHF fields of low (nonthermal) intensity (10 mW/cm) in three frequency ranges — millimeter, centimeter (10-cm), and decimeter (Lobanova, 1959, 1964b; Lobanova and Tolgskaya, 1960).

After the elaboration of a positive and differential reflex the animals were irradiated daily for 30-60 min and their conditioned-reflex activity was assessed from the length of the latent period (the interval between the signal and the motor reaction) and from the percentage of failures of the differential reaction.

Irradiation with millimeter waves had a relatively weak effect: There was a slight reduction in the latent period (to light and sound stimuli) and failure of the differential reaction (in 62% cases), but these effects did not occur until after 48-52 exposures.

Irradiation with centimeter waves (36 exposures in all) had a much greater effect. The first few exposures led to failures of the positive conditioned reflexes and by the end of the treatment the latent period had increased by a factor of 3-4 (from 2 to 7 sec).

The differentiation reactions, however, were unaffected. The animals recovered in eight weeks.

After 50-52 exposures to decimeter waves there was some reduction in the latent period (to 17-74% of the control value) and differentiation was upset in 50% cases. After the next 54 exposures opposite changes occurred. In all the described experiments there was an increase in excitability and a weakening of active inhibition in the first period of irradiation and a reduction of excitability and even the development of prohibitive inhibition in the second period. Normal reactions were reestablished 3-8 weeks after the end of the radiation treatment.

Investigations with SHF fields of higher (thermal) intensities revealed essentially the same effects. Exposure of rats to 10-centimeter waves with an intensity of 16-94 mW/cm^2 for 1 min per day led at first to enhanced excitability (after 14-20 exposures) and then to reduced excitability (Minecki et al., 1962; Minecki and Romanik, 1963). Exposure of rats to a field of wavelength 1.25 cm and an intensity of 109 mW/cm^2 for 15-30 min per day led to an increase in the latent period after only three exposures (Tallarico and Ketchum, 1959). A single exposure of mice to 3-centimeter waves with an intensity of 400 mW/cm^2 for 5 min led to failure of conditioned reflexes in 28% cases and to failure of differentiation in 68% cases (Gorodetskaya, 1960, 1964a).

Experiments with dogs revealed a much more complex variation of the effects of EmFs in relation to the exposure conditions and the type of higher nervous activity of the animals. Firstly, exposure to low intensities causes more significant changes in conditioned-reflex activity than exposures to high intensities. Secondly, irradiation of the head region is more effective than irradiation of the whole side of the body. Thirdly, while irradiation of animals with the strong type of higher nervous activity leads to definite disturbances in conditioned-reflex activity, in animals of the weak type these disturbances are of an indefinite nature or there is a disturbance of the entire higher nervous activity. Fourthly, considerable changes occur in unconditioned-reflex activity, as well as in conditioned-reflex activity. To illustrate these features we cite the results of some investigations.

The temporal, frontal, and occipital regions of the head of dogs were exposed to a high-intensity UHF field (50 MHz) and the changes in conditioned-reflex activity were assessed from the rate of salivation and the failure of differential reflexes (Livshits, 1957a, 1957b, 1958). In animals with the strong type of nervous system the reflex was enhanced by a factor of 2-3 and differentiation was destroyed, whereas in animals with the weak type of nervous system salivation was increased, but differentiation was adversely affected. These effects were often produced by relatively low intensities (7-12 W on the electrodes) and were absent at higher intensities (20-55 W).

Exposure of the side of the body of dogs to intense SHF fields (100-200 mW/cm^2) (Subbota, 1957b, 1958) reduced salivation and almost tripled the latent period of this reaction, but differentiation was not affected. Such effects were clearly manifested in animals of the strong type and were indefinite in animals of the weak type. Exposure to low intensities (5-10 mW/cm^2, 1 h) led to an enhancement of the reflex and an increase (of up to 24%) in the latent period in animals of the strong balanced type, but to a reduction in both of these indices in animals of the strong unbalaced type. Repeated exposures to the same intensity (for 2 h per day) led after 18 exposures to an increase in salivation and an increase in the latent period in animals of the strong type.

The described experiments show that SHF fields of high intensity adversely affect the conditioned-reflex activity of dogs, whereas low intensities stimulate such activity. However, experiments with even lower intensities led to a different result (Svetlova, 1962; cited by Lobanova, 1964b): Exposure of the side of the body of dogs to decimeter waves with an intensity of only 2 mW/cm^2 led to the suppression of the reflex and an increase in the latent period; multiple exposures to an intensity of 0.2-0.3 mW/cm^2 also had a suppressing effect.

A stimulating effect of low-intensity EmFs has also been found at high and ultrahigh frequencies (Salei, 1964). Exposure of the head of dogs to EmFs of frequencies 27, 30, 39, 94, 102, and 120 MHz and intensity of 3 V/cm for 20 min led to a 35-65% increase in unconditioned-reflex salivation. The protein content of the saliva was altered to an even greater extent (by 87-108%). These functional changes in salivary-gland activity were more persistent after exposure to a field of 3 V/cm than to a field of higher strength (25 V/cm), which raises the temperature of the salivary gland tissues.

Of considerable interest are the investigations (Kitsovskaya, 1960, 1964a) of the effect of pulsed SHF fields on the special unconditioned reaction of a specially bred population of rats, which respond to the sound of a bell by a sudden motor reaction or a convulsive fit.*

Groups of animals were exposed to decimeter, 10-centimeter, 3-centimeter, and millimeter waves with intensity 10 mW/cm² for 1 h per day. The changes in the reactions to sound were assessed as the exposures proceeded and the percentage of cases in which changes occurred were observed. All these kinds of treatment had a similar effect — reduction of the sensitivity of the animals to sound stimulation. However, the degree of this effect decreased with reduction in the wavelength.

One to four exposures to decimeter waves led to considerable weakening of the reaction: In 83% of the cases the two-wave reaction, terminating in a fit, was converted to a motor reaction without a fit, or to no reaction at all; a single-wave reaction ending in a fit was converted to a two-wave reaction without a fit.

Ten-centimeter waves began to produce an effect after 23-25 exposures, but in 100% of the cases the single-wave reaction ending in a fit was converted to a two-wave reaction and the animals which gave a two-wave motor reaction ceased to react to sound at all after irradiation.

The effect of 3-centimeter waves was manifested after 37-40 exposures: In 58% of the cases the single-wave and two-wave reactions ending in a fit persisted, but the fits were not so severe.

Millimeter waves also caused a weakening of the reactions in 75% of the cases after 48 exposures. This, however, was manifested mainly in an increase in the latent period of the reaction, a prolongation of the first wave of excitation, and a shortening of the inhibition period.

---

*These animals show a first wave of excitation (sudden movements) 5-15 sec after the bell is switched on. This is followed by a defense reaction in the form of an inhibitive process — a state of rest for 10-20 sec. Finally, there is a second wave of excitation, which may terminate in a convulsive fit. When the inhibitive process is weak the first wave of excitation may terminate in a fit.

The relationship between such effects and the irradiation intensity in the 10-centimeter range was investigated. In the case of exposures for 1.5 h to an intensity of 1 mW/cm^2 the changes were manifested after 23-25 exposures, but their number reached a maximum (83% of the cases) after 117 exposures.

The discussed experimental data indicate fairly definitely that the effect of EmFs on conditioned and unconditioned reflexes is due to their action on the central nervous system. We can now point out some of the characteristic features of this action.

It is obvious that EmFs either have a reflex action on the central nervous system — through the peripheral elements of the nervous system, or they act directly on the brain structures. The first feature of the action of EmFs is that the nature of the effect produced depends on the site of action of the EmF. This can be seen by comparing the nature and degree of the observed effects of EmFs of different frequency ranges with the corresponding depth of penetration of the energy into the body tissues (see Section 4.3). A constant magnetic field and EmFs up to the UHF range penetrate the brain structures of both small and large animals, whereas decimeter and 10-centimeter waves reach the brain only of small animals. We have seen that in such cases all these fields produce similar effects — either conditioned stimulation, a change in conditioned-reflex activity, or a change in unconditioned reflexes. These effects are much less pronounced (and sometimes opposite in nature) in the case of EmFs in the 3-centimeter range and, in particular, in the millimeter range, where practically all the energy is absorbed in the skin layer. In addition, we have seen that exposure of the head causes much greater effects than whole-body exposure.

The second characteristic feature is the special relationship between the effect of EmFs on the central nervous system and the field strength. In some cases a lower intensity leads to more pronounced effects (change in conditioned reflexes in dogs due to UHF irradiation). In other cases, opposite effects alternately occur as the intensity is reduced; a phase of inhibition of conditioned reflexes at intensities of 100-200 mW/cm^2, a stimulating phase at 5-10 mW/cm^2, and another phase of inhibition at 0.2-2 mW/cm^2.

The third feature is the cumulative nature of multiple exposures to low-intensity EmFs. Here we encounter opposite phases —

an increase in excitability of the central nervous system after the first few exposures and a decrease in excitability after subsequent exposures.

Finally, the fourth feature is that direct exposure of the brain to EmFs has a stimulating effect when the EmF is used as a conditioned stimulus and in the case of unconditioned reflexes, but the effect on previously elaborated conditioned reflexes is of an inhibitive nature.

7.3. Effect of EmFs on Nervous Regulation

of Cardiovascular System

In the study of the effect of EmFs on the control of physiological processes the action of EmFs in the high-frequency to superhigh-frequency ranges on cardiovascular function has been most convincingly demonstrated. A considerable amount of the data has been obtained from clinical investigation of persons who have been chronically exposed to EmFs; the most extensive work of this kind has been carried out in the USSR (Orlova, 1960a, 1960b; Sadchikova, 1964; Konchalovskaya et al., 1964; and others). Table 11 gives typical results of such investigations (Sadchikova, 1964).

As the table shows, chronic exposure to EmFs of various frequencies leads to similar changes in cardiovascular function — reduction of blood pressure, heart rate, and intraventricular conductivity. These changes are most pronounced, however, in the case of exposure to SHF, when the electromagnetic energy is absorbed in the surface tissues of the human body. Hence, we can postulate that these changes are due to the direct action of EmFs on the surface receptors (either to their stimultation or to alteration of their functional state). This hypothesis is consistent with our knowledge of the physiology of nervous regulation of the functions of the cardiovascular system. It is known that this set of changes is typical of the vagotonic changes in vegetative neural regulation produced by the action of very diverse stimuli on the surface receptor regions (Ginetsinskii and Lebedinskii, 1956).

Experimental investigations with animals have confirmed the vagotonic nature of the action of EmFs. They have also revealed other special features of the reactions of the cardiovascular system to EmFs and have shown some aspects of the way in which these

TABLE 11. Disturbances of Cardiovascular Function in Persons Chronically Exposed to EmFs of Various Frequencies

EmF parameters		Ratio of percentage of cases with particular defect due to EmF to percentage of cases in control (not exposed to EmF)		
range	intensity	reduced blood pressure (arterial hypotonia)	slow heart beat (bradycardia)	QRS interval in ECG increased to 0.1 sec (reduced ventricular conductivity)
SHF (centimeter waves)	From one to several mW/cm^2	1.85	24	11.5
UHF	< 1 mW/cm^2	2.0	16	12.5
	Low, not thermal	1.2	8	21
Short-wave HF	Tens to hundreds of V/m	0.21	12	—
Medium-wave HF	Hundreds to 1000 V/m	1.2	5	—
Percentage of cases in control		14%	3%	2%

reactions depend on the EmF parameters and the region of body exposed.

The effect of chronic exposure to pulsed SHF fields on the blood pressure was investigated in experiments with rats (Gordon, 1964b). The animals (220 in all) were divided into four experimental groups and a control group. They were subjected respectively to irradiation with waves of the decimeter, 10-centimeter, 3-centimeter, and millimeter ranges. The irradiation intensity for all

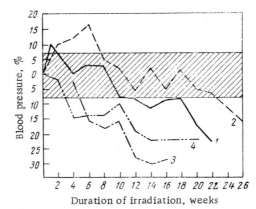

Fig. 39. Change in blood pressure of rats due to chronic exposure (1 h per day) to low-intensity SHF fields. 1) Decimeter waves; 2) 10-centimeter waves; 3) 3-centimeter waves; 4) millimeter waves. Hatched region — control.

the groups was 10 mW/cm^2 and the duration of exposure was 1 h per day. The course of irradiation lasted 6-8 months.

Figure 39 shows plots of the change in blood pressure against the number of exposures. The figure shows that in the case of decimeter and 10-centimeter waves the changes were of a two-phase nature — an increase in the first few weeks of irradiation and a subsequent reduction from the 20th-24th week. In the case of 3-centimeter and millimeter waves there was only one phase — a reduction of pressure, which began within the first few weeks of irradiation. A comparison of the effect of pulsed and continuous 10-centimeter waves showed that their ultimate effect was the same. In the case of continuous waves, however, the changes appeared much earlier (in the 8th week of irradiation) and consisted of a single phase. In all cases the pressure gradually (within 8-10 weeks) returned to normal after the cessation of irradiation.

Differences between the effects of the long-wave and short-wave ranges of SHF fields were manifested in the variation of the magnitude of the effect with the irradiation intensity. Table 12 shows that 3-centimeter waves with an intensity of 1 mW/cm^2 had practically no effect and that an increase in intensity from 10 to 40-100 mW/cm^2 did not alter the effect of 3-centimeter and millimet-

TABLE 12. Reduction of Blood Pressure in Rats Due to Chronic Exposure to SHF Fields of Different Frequency Ranges and Intensities

Frequency range	Irradiation intensity, mW/cm^2	Blood pressure reduction in comparison with control, mm Hg
Millimeter	10	18.7 ± 3.53
	40	19.14 ± 5.81
3-centimeter	1	9.2 ± 8.8
	10	23.0 ± 5.01
	100	25.0 ± 5.94
10-centimeter	1	14.7 ± 4.4
	10	9.4 ± 2.3
Decimeter	1	7.2 ± 2.2
	10	17.5 ± 6.06

er waves. Decimeter and 10-centimeter waves were effective at an intensity of 1 mW/cm^2. It is particularly remarkable that the effect due to 10-centimeter waves was reduced (!) when the intensity was increased by a factor of 10.

Other investigations (Nikonova, 1964a) have shown an effect of high-frequency EmFs (500 kHz) on the blood pressure of rats. For 2 h daily over a period of 10 months one group of animals was placed in the region where the electric component was dominant (1800 V/m) and another group was placed in the region of the magnetic component (50 A/m). In the first group the pressure began to fall after 5 months of treatment and reached a level 5% below the control value (P < 0.1), while in the second group it began to fall in the 7th month and reached a level 10% below the control value (P < 0.01). Thus, the change consisted of a single phase.

The investigations discussed indicate that the nature and magnitude of the effect of EmFs on cardiovascular function depend mainly on whether the peripheral or central nervous system is irradiated. More definite information on this question, however, was obtained in investigations of the heart rate in rabbits when different regions of the body were exposed to SHF fields (Presman and Levitina, 1962a, 1962b; Levitina, 1964, 1966b).

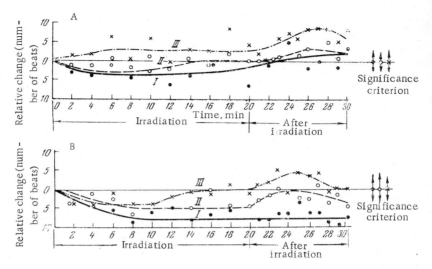

Fig. 40. Change in heart rate of rabbits in continuous SHF fields of low intensity. A) Irradiation of back of body; B) irradiation of underside of body. I) Whole-body irradiation; II) irradiation of back (in case A) or belly (in case B); III) irradiation of head.

In the first series of experiments the animals were irradiated with pulsed (1 μsec, 700 pulses/sec) and continuous 10-centimeter waves at intensities of 7-12 and 3-5 mW/cm^2 (mean power), respectively. The results of the experiments are represented graphically in Figs. 40 and 41, which show that irradiation of the underside of the body reduced the heart rate (negative chronotropic effect), whereas irradiation of the back of the body or both sides of the head had the opposite effect, i.e., acceleration of the heart rate (positive chronotropic effect). Two special features were noted: First, the negative chronotropic effect occurred immediately after the start of irradiation; second, pulsed irradiation had a greater effect than continuous irradiation, although the intensity in the second case was a little higher.

These results suggested that the negative chronotropic effect of SHF fields is due to direct action on the peripheral nervous system and the positive effect is due to action on the central nervous system. This was confirmed in a subsequent series of experiments in which different regions of the body were exposed to more intense

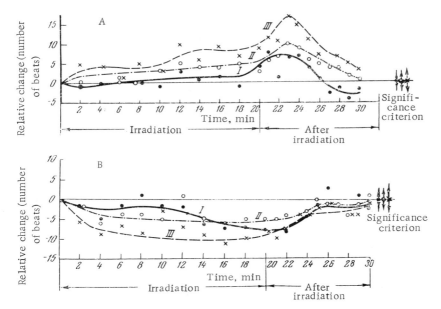

Fig. 41. Change in heart rate in rabbits in pulsed SHF fields of low intensity. A) Irradiation of back of body; B) irradiation of underside of body. Other symbols as in Fig. 40.

radiation — pulses of continuous waves (700–1200 mW/cm^2 per pulse) or a series of short pulses (1 μsec long, 700 pulses/sec, mean power in series 350–380 mW/cm^2), the duration of the pulses and series being 0.1 sec and the repetition rate two per second. The results of the experiments were unexpected: Irradiation of any region of the body, including the head, led only to the negative chronotropic effect. Irradiation of the same regions of the body with the skin anesthetized (by a trimecain solution) did not lead to any changes in the heart rate.

A comparison of these results with those of the previous series of experiments showed that: 1) the direct action of SHF fields on the ventral surface receptor regions produces the negative chronotropic effect, irrespective of the intensity, but this effect is produced only by relatively high intensities when the back of the body and, in particular, the head are irradiated; 2) the positive chronotropic effect is only produced by the direct action of low-intensity SHF fields on the central nervous system and has a long latent period.

In subsequent experiments a rough estimate of the relationship between the positive chronotropic effect and the irradiation intensity was made (accurate dosimetry was too difficult). It was found that in the case of irradiation of the head the effect occurred at intensities of a few mW/cm^2, whereas at higher intensities (tens of mW/cm^2 and more) the negative chronotropic effect was observed. Finally, quite low intensities (less than 1 mW/cm^2) caused no change in the heart rate.

Special experiments with frogs were undertaken to obtain conclusive evidence that the chronotropic effect of SHF fields is due to their action on different parts of the nervous system (Levitina, 1966a). The negative chronotropic effect was produced by irradiation of the animal's body with an intensity of only 0.06 mW/cm^2, while the same irradiation of the head produced the positive chronotropic effect. When the nervous regulation of the heart was interfered with in any way (by anesthesia, transection of the cardiac nerves) the heart rate was unaffected by irradiation. No effect was produced by direct irradiation of the denervated heart in situ.

Vagotonic changes due to a constant magnetic field have also been observed (Knepton and Beischer, 1964). Exposure of the head of a monkey to a field of 20,000–70,000 G or of the heart region to a field of 625 G for 40–180 min reduced the heart rate from 325 to 225, increased the degree of sinus arrhythmia, and altered the electrocardiogram. On cessation of irradiation the heart rate soon returned to normal.

A similar effect has been observed (Cassiano et al., 1966) in the case of exposure of man to a magnetic field (33.8 G for 1 h): The pulse rate was reduced by 15% (P = 0.05).

Interesting data regarding the vagotonic effect of SHF fields have been obtained in investigations of cholinesterase activity in the blood and organs of animals subjected to chronic exposure to decimeter, 10-centimeter, or millimeter waves of equal intensity (10 mW/cm^2) for 1 h per day (Nikogosyan, 1960, 1964b). The cholinesterase activity in the blood of rabbits and rats and in different organs of rats was determined. It was found that all these kinds

TABLE 13. Comparative Data on Reduction of Cholinesterase Activity in Blood and Organs of Animals Due to Chronic Exposure to SHF Fields

Frequency range	Percentage of animals showing changes	Blood serum			Brain stem			Liver and heart		
		number of exposures		percentage of changes in relation to control	number of exposures		percentage of changes in relation to control	number of exposures		percentage of changes in relation to control
		before appearance of first changes	before appearance of marked changes		before appearance of first changes	before appearance of marked changes		before appearance of first changes	before appearance of marked changes	
Decimeter	100	10-20	100 10-centimeter	36	28	100 10-centimeter	48	100	100	12-34
10-centimeter	70	30-40	40	27	76	100	29-30	140	140	0-50
Millimeter	63	180	180	15	180	180	37	No changes after 180 exposures		

Note. The cholinesterase activity in the cerebral hemispheres was not affected even by 180 exposures to fields of any of the frequencies.

of irradiation had a similar result — the reduction of cholinesterase activity. As Table 13 shows, the magnitude of this effect decreased with reduction in the wavelength. The changes produced by any of the wavelengths were greatest in the brain stem, moderate in the blood, and slight in the liver and heart. There was no change at all in the cerebral hemispheres. The change in cholinesterase activity in the blood and organs due to chronic exposure to SHF fields of much higher intensity (100 mW/cm^2) was of a different nature: After the first few exposures the activity was enhanced, but subsequently (after 6-12 exposures) it was reduced. In every case the cholinesterase activity returned to normal 40-45 days after the cessation of irradiation.

We again encounter here the similarity of the effects of SHF fields of different frequencies, as was found in the case of the action of these fields on the functions of the central and vegetative nervous system. There is a difference in the magnitude (but not in the nature) of the effect produced by waves absorbed in the skin layer (millimeter waves) and those which penetrate into the deep-lying tissues.

Thus, the action of EmFs on the cardiovascular system reveals the same features as in the previously described disturbances of the regulatory functions of the nervous system, viz., the cumulative effect of exposures to weak fields, the frequency-independent nature of the effect, the variation of the magnitude of the effect with the frequency (due mainly to the depth of penetration of the EmF energy), and the difference in the nature of the effects due to direct action on the peripheral and central structures of the nervous system. Special attention should be drawn to the peculiar way in which the nature and magnitude of the effects vary with the intensity of EmFs acting directly on the central nervous system: Firstly, the effects produced by low and high intensities may be of opposite nature; secondly, the effects produced by low intensity may be more pronounced than that produced by high intensity, and, thirdly, an intensity exceeding certain small values may have no effect at all. On the other hand, the effects due to the direct action of EmFs on the peripheral receptors are practically independent of the EmF intensity in a fairly wide range.

As we have seen, effects of EmFs on the central nervous system have also been discovered by observation of the behavior of animals, by the investigation of conditioned and unconditioned reflexes, and by an investigation of cardiovascular reactions. Several investigations carried out in recent years, however, have shown direct reactions of the brain structures to EmFs. These investigations have been conducted along two main lines: studies of the effect of EmFs on the electric activity of brain structures and on the sensitivity of different parts of the central nervous system to various stimuli.

7.4. Effect of EmFs on Electric Activity of Brain and on Sensitivity of Central Nervous System to Other Stimuli

Investigations of the effect of EmFs on the electroencephalogram (EEG) of animals and man have been described in Kholodov's recently published monograph (1966), which contains an extensive bibliography. Hence, in this section we will cite only the summarized results of experimental investigations and consider some particular studies which have a special bearing on the general problem discussed in this book — the problem of the biological activity of EmFs.

Electroencephalographic investigations have involved exposure of the head of animals, individual regions of the body and, finally, the whole body (usually one side) to EmFs.

The EEG has been recorded before, during, and after exposure to the EmF, but in some experiments (owing to interference on the electrodes) the EEG has been recorded only before and after irradiation. Investigations have been conducted on the intact (undamaged) brain, on brain in which certain subcortical structures have been injured (by electrocoagulation), on isolated brain (transected at the level of the midbrain) and, finally, on a neuronally isolated strip of the cerebral cortex.

The main result of the numerous experimental investigations is that EmFs of a very wide range of frequencies and intensities affect the electric activity of the cerebral cortex and subcortical structures. This effect is manifested in the induction of the following types of changes in the EEG·

a) increased synchronization — an increase in the number of slow high-amplitude waves and "spindles." These changes occur on the elapse of a considerable period of time (latent period) — tens and even hundreds of seconds — after the start of irradiation;

b) prolonged desynchronization — a reduction in the amplitude of the main rhythm of biopotentials and an increase in the number of high-frequency waves. These changes also appear after a long latent period;

c) short-term desynchronization, which occurs soon after application and removal of the field;

d) an aftereffect similar to enhanced desynchronization, but occurring a fairly long time after the end of irradiation;

e) the occurrence of convulsive epileptoid discharges (of high frequency and amplitude).

Figure 42 shows electroencephalograms illustrating these kinds of changes.

The changes are quantitatively assessed by three main quantities: "the strength of the reaction" (number of cases of changes as a percentage of the total number of exposures), the length of the latent period (in seconds), and the increase or reduction in the amplitude of the biopotentials (in mW or %).

A comparison of the results of investigations (conducted mainly on rabbits) under various experimental conditions reveals some typical features of the way in which the nature and magnitude of the changes in the EEG vary with the frequency and intensity of the EmF, the region of the body exposed, and the functional state of the irradiated structures.

The changes in the EEG due to exposure of an animal's head to EmFs of various frequencies (from hundreds of kHz to 2000–3000 MHz) and to a constant magnetic field are of the same nature. The main reaction produced by these forms of treatment is an increase in the number of slow high-amplitude oscillations and the number of spindles in the biopotentials of the cerebral cortex. In some cases there is also a repetition of this reaction after the end of irradiation and, finally, there are the short-term desynchronization reactions on application and removal of the field. Alteration of the frequency leads only to some variations in the main reaction: In the UHF range the slow high-amplitude oscillations are most pronounced and the appearance of spindles is rarer. In the SHF range these two kinds of changes are equally pronounced. In the case of a constant magnetic field the main change is the appearance of spindles (Chizhenkova, 1966).

The main reaction is also of the same type in a wide range of EmF intensities (Nikonova, 1965b; Zenina, 1964; Gvozdikova et al., 1964; Kholodov, 1966; Nizhenkova, 1966): from 50 to 1000

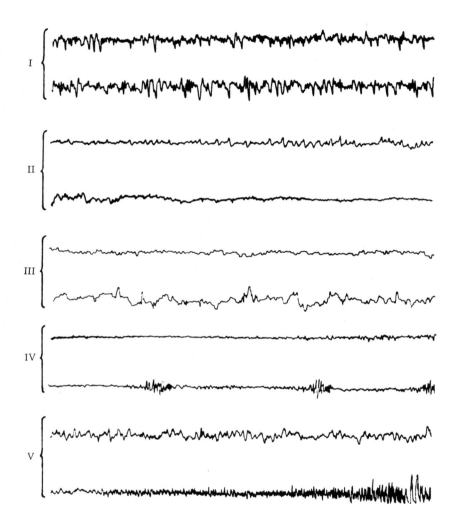

Fig. 42. Characteristic changes in EEG of rabbit due to exposure to EmFs of various frequencies and a constant magnetic field. I) Increase in amplitude of biopotentials of main rhythm (synchronization); II) reduction of amplitude of biopotentials of main rhythm (desynchronization); III) increase in number of slow waves; IV) appearance of "spindles"; V) appearance of convulsive (epileptoid) discharges.

V/m for high-frequency EmFs, from 30 to 5000 V/m fields, from 0.2 to 1000 mW/cm^2 in the UHF range, and from 200 to 1000 Oe for a constant magnetic field.

Exposure of other parts of the rabbit's body as well as the head to intense EmFs of different frequencies leads to another set of changes in the EEG in addition to the main reaction. These changes are a reduction in the amplitude of the biopotentials (desynchronization) and an increase in the number of high-frequency oscillations. Such changes have been observed as a result of exposure to high-intensity EmFs of low, high, and ultrahigh frequencies (Shmelev, 1964a; Kholodov, 1966). In the case of SHF irradiation of the side of an animal's body such reactions may predominate over the main reaction even in the case of low intensities — of the order of 2-10 mW/cm^2 (Gvozdikova et al., 1964). Finally, such reactions have been observed as a result of direct action on the peripheral nerves: Exposure of the sciatic nerve of a cat to 3-centimeter waves of high intensity leads to a reduction in the amplitude of the biopotentials, as Fig. 43 illustrates (Fleming et al., 1961). Exposure of the back of a rat to intense 1.25-cm waves, which are completely absorbed in the skin, leads to high-frequency oscillations of the biopotentials 7 min after irradiation (Keplinger, 1958).

The quantities characterizing the electric reactions of the rabbit brain to EmFs are practically independent of the EmF frequency at comparable intensities. On exposure to a constant magnetic field of 200-1000 Oe the strength of the main reaction is 37-52%, and on exposure to EmFs of high to superhigh frequencies (30-

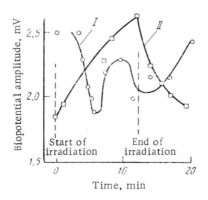

Fig. 43. Change of biopotential amplitudes in rabbit brain during irradiation of sartorius with 3-cm waves. I) Amplitude of biopotentials; II) nerve temperature.

1000 V/m, 2-10 mW/cm^2) it is 40-49%; the latent periods on exposure to a constant magnetic field vary in the range 24-55 sec,
and on exposure to EHF and SHF fields they are 30-40 sec. The
amplitude of the biopotentials is increased by UHF and SHF fields
up to 500-700 μV (these data are taken from the above-cited works
of Soviet authors).

The values of these indices depend considerably on the intensity of the EmF. The nature of this dependence is different for the
main reaction and for the reaction involving desynchronization and
increase in the number of high-frequency oscillations. The results
obtained on exposure of the side of a rabbit's body to SHF fields
(decimeter and centimeter waves) (Zenina, 1964; Gvozdikova et al.,
1964; Kholodov, 1966), which are graphically represented in Fig.
44, illustrate this. The graphs show that at thermal intensities the
strength of the main reaction diminishes with intensity reduction,
but at lower intensities it is practically independent of the intensity
and even shows a tendency to increase with intensity reduction.
Desynchronization accompanied by an increase in the high-frequency
oscillations occurs only at intensities above 2 mW/cm^2. The latent
periods of the two reactions become longer with intensity reduction,
but the period of the main reaction is shorter and the curve of its
variation has a smaller gradient than for the second type of reaction.

Interesting data on changes in the electric activity of the
brain due to a constant magnetic field have been obtained in experiments with lizards (Becker, 1963b). The EEG was recorded continuously during exposure to the field, which gradually increased
from 500 to 3800 G and then decreased to its initial value. The
amplitude of the slow oscillations gradually increased with increase
in field strength and also when the strength was reduced.

In a series of experiments Kholodov and Zenina (1964), Luk'-
yanova (1965), Kholodov (1966), and Chizhenkova (1966) compared
the effect of EmFs on the electric activity of the brain and in isolated brain structures of rabbits, and also the change in their reactions on destruction of particular brain structures and on enhancement of their excitability by the injection of caffeine or adrenalin. The following general features were discovered:

1. The nature of the changes in the EEG (recorded from the
cerebral cortex) due to a constant magnetic field, UHF field, or

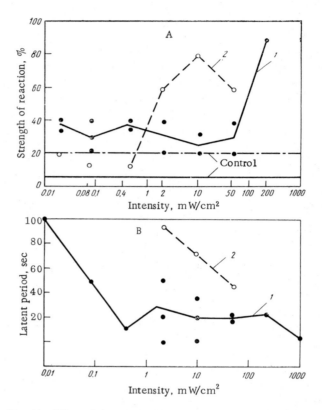

Fig. 44. Effect of decimeter and centimeter waves on rabbit
EEG in relation to irradiation intensity. A) Change in strength
of reaction; B) change in latent period; 1) increase in slow
high-amplitude waves and spindles; 2) reduction of amplitude
of biopotentials of main rhythm and increase in high-frequency
waves. The graphs are drawn from the data obtained by Zenina (1964), Gvozdikova et al. (1964), and Kholodov (1966).

SHF field is not altered by destruction of the visual, auditory, or
olfactory analyzers, or by injury to the hypothalamus, thalamus,
or reticular formation of the midbrain.

2. Reactions of the isolated brain (transected at the level of
the midbrain) and a neuronally isolated strip of cortex to these
fields are of the same nature as those of the undamaged brain.
The reaction of these isolated preparations is more pronounced

than that of the intact brain — the strength of the reaction is greater and its latent periods are shorter.

3. Injection of caffeine or adrenalin into animals enhances the main reaction and convulsive discharges of epileptoid type appear in the EEG. The effect of caffeine injection is even more pronounced when the EEG is recorded from the isolated brain.

4. The nature and time of the main reaction to EmFs are the same in all parts of the brain. There is evidence that the hypothalamus and cerebral cortex are more sensitive to a constant magnetic field than the thalamus and reticular formation of the midbrain. The enhancement of the reaction to EmFs by the injection of exciting substances is correlated with the sensitivity of the brain structures to them: Caffeine enhances the cortical reaction most, while adrenalin has most effect on the reactions of the hypothalamus and reticular formation.

The effect of EmFs on the sensitivity of the central nervous system to other stimuli (Shmelev, 1964b; Kholodov, 1966; Chizhenkova, 1966) has also been investigated. It was found that a constant magnetic field reduces the sensitivity of mammals to photic stimulation, amphibians to chemical stimulation (Turk reflex), and fish to electric stimulation; UHF and SHF fields, on the other hand, increase the sensitivity of rabbits to photic stimulation.

An effect of SHF fields on the electric activity of the human brain is indicated by the results of investigations of large groups of persons which have been chronically exposed to low-intensity fields (Drogichina et al., 1962; Ginzburg and Sadchikova, 1964; Klimkova-Deicheva and Rot, 1963; Sinisi, 1954; Sarel et al., 1961, and others). These investigations have shown that in people who have worked for a long time in SHF fields the number of slow high-amplitude waves in the EEG is slightly higher than normal. There is also a reduction in the sensitivity of the olfactory analyzer in people working with SHF generators (Lobanova and Gordon, 1960) and with short-wave and ultrashort-wave generators (Fukalova, 1964c).

Thus, we can regard it as experimentally established that EmFs affect the electric activity of various parts of the brain. The nature of the effect due to EmFs of a very wide range of frequencies — from a constant magnetic field to SHF fields — is the. same, but it depends on the part of the body irradiated and the EmF intensity. When brain structures are directly exposed to EmFs there is a fairly long latent period, after which the number of slow high-amplitude oscillations and spindles increases. Such changes, usually observed during sleep and anesthesia, appear to indicate that EmFs have an inhibiting effect on brain structures. It is characteristic that such reactions to EmFs are produced by either high- or low-intensity fields and may even become slightly more pronounced when the EmF intensity is reduced. If the peripheral, as well as the central, nervous structures are exposed to an EmF the changes produced — the predominance of low-amplitude oscillations in the electroencephalogram — are similar to those which are characteristic of arousal reactions and other reactions associated with stimulation of the central nervous system. It is characteristic that these reactions to EmFs are produced not only by fields of fairly high intensity, but also of low intensity.

The discussed experimental results indicate that the cerebral cortex and the interbrain structures, particularly the hypothalamus, are very sensitive to EmFs. We know, however (Drischel, 1956; Magoun, 1958; MacIlwain, 1959; Brady, 1958), that these structures are responsible for the central regulation of physiological processes, which, as we have already described, are affected by EmFs. The hypothalamus is responsible for humoral regulation of the blood sugar content and the composition of the formed elements of the blood; it regulates the activity of the endocrine organs — the hypophysis, the adrenal cortex, and so on. Direct electric stimulation of the hypothalamus causes changes in the number of eosinophils and lymphocytes, in the secretion of the adrenocorticotrophic hormone by the hypophysis, in the ascorbic acid content of the adrenals, and so on.

Of interest in this connection is the indirect evidence, obtained in investigations of the humoral changes in the body of animals exposed to EmFs, that EmFs affect the central nervous system.

7.5.   Effect on Humoral Regulation

Several investigations have revealed various manifestations of the action of EmFs on the hemopoietic function, the formed elements (cells), and the protein and mineral composition of the blood.

The change in the percentage composition of the formed elements of the blood has been investigated in a great variety of conditions — single and chronic exposures to high- and low-intensity
EmFs.

In the above-described investigations on dogs (Howland et al.,
1961; Michaelson et al., 1961a, 1964) a single 6- to 8-h exposure
to high-intensity SHF fields (2800 MHz, 100 mW/cm^2 and 200 MHz,
165 mW/cm^2) led to a 25-55% increase in the number of leucocytes
24 h after irradiation, whereas the numbers of lymphocytes and eosinophils fell immediately after irradiation, but became higher than
normal after 24 h. The percentage hemoglobin content varied in
the same way. In other experiments on dogs (Tyagin, 1957) a similar exposure (100-300 mW/cm^2) for 25 min led to a two-phase
change in the number of leucocytes — a reduction immediately after irradiation and an increase after 2-4 h.

Exposure of rats to shorter waves (10,000 MHz) with an intensity of 400 mW/cm^2 for 5 min led to a reduction in the numbers
of leucocytes and erythrocytes, an effect which persisted for five
days after the end of irradiation (Gorodetskaya, 1960; 1964a).
Exposure of rats to even shorter waves (24,000 MHz) with an intensity of 20 mW/cm^2 for 7.5 h led to an increase in the number of
erythrocytes and the hemoglobin content, but to a reduction in the
number of leucocytes, immediately after exposure. After 16 h,
however, all the blood indices began to increase slowly and reached
the normal level again in 14 days (Deichmann et al., 1959, 1964).
Chronic exposure of rats to 3000-MHz fields with an intensity of
40-100 mW/cm^2 (Kitsovskaya, 1964b) led to a reduction in the number of leucocytes, but did not affect the percentage hemoglobin content.

A constant magnetic field also affects the blood of animals
which have been chronically exposed (Barnothy and Barnothy, 1960).
In rats exposed to a field of 4200 Oe the number of leucocytes fell
(by 30-40%) during the first 10 days of treatment and became 100%
higher than in the control in the following 2-3 weeks.

The described effects show a great deal in common; first,
the two-phase nature of the changes; second, the similarity of
the changes due to EmFs of different frequencies, and thirdly, the
fairly rapid reversibility of the changes — the return to normal

of the blood indices within a period of a few hours to several days. At the same time, there is considerable variation in the nature and magnitude of the changes depending on the species, and even strain, of animal. For instance, in the above-described investigations of Deichmann, exposure of rats in the same conditions led to an increase in the percentage hemoglobin content in rats of the Osborne-Mendel strain, but to a decrease in rats of the Fisher strain.

A change in the blood indices in people who have been exposed for a long time (from one year to several years) to weak EmFs of various frequencies has been reported. A considerable number of investigations have revealed changes of the same type — an increase in the number of formed elements and the percentage hemoglobin content due to the action of EmFs in the medium-wave, short-wave, ultrashort-wave, decimeter, centimeter, and millimeter ranges (Sokolov and Chulina, 1964). Altered protein composition of the blood is also typical of people who have been chronically exposed to radio-frequency EmFs of low intensity (Gel'fon and Sadchikova, 1964). In 50% of the investigated subjects the total protein content was increased — mainly due to the increase in the amount of globulins, as was indicated by the change in the albumin-globulin ratio.

Another characteristic change was the increase in the histamine content of the blood. Such changes were observed in people exposed to EmFs of various frequencies — from the medium-wave to the centimeter range. Shorter waves, however, were more effective. Changes in the histamine content due to SHF fields were discovered in experiments with animals (Gel'fon, 1964). In rabbits exposed to 10-centimeter waves with an intensity of 10 mW/cm^2 for 1 h per day the histamine content varied in the first 5 months, periodically increasing and decreasing, but it remained above normal all the time.

Thus, the changes in the protein composition and histamine content of the blood due to EmFs have the above-mentioned common features — similar nature irrespective of EmF frequency, two-phase variation, and rapid reversibility.

The effect of SHF fields on the mineral composition of the blood has been investigated in experiments involving chronic exposure of rats (Kulakova, 1964). Animals exposed to decimeter waves with an intensity of 40 mW/cm^2 for 1 h per day showed an

increase in the Ca-ion content of the blood plasma after six exposures, but there was a much greater increase in their amount in
the urine after 17 exposures. During this time the Na- and K-ion
content was unaltered. The change in Ca content affected the specialized forms of appetite of animals which had been exposed to
centimeter and decimeter waves in the above indicated conditions.
Of the proferred solutions of NaCl, KCl, and $CaCl_2$ (in 20% glucose
solution) the animals consumed a much greater amount of the calcium solution. Such behavior was observed in ordinary rats and
rats of the Wistar strain after 5-7 exposures to 10-centimeter
waves with intensities of 10 and 40 mW/cm^2. In the case of exposure to decimeter waves with an intensity of 40 mW/cm^2, 26 exposures were required before rats of the Wistar strain showed an
enhanced appetite for the calcium solution, and exposures to an intensity of 10 mW/cm^2 had no effect.

The metabolism of animal tissues and organs is affected by
EmFs of various frequency ranges — from low to superhigh.

The effect of EmFs on carbohydrate metabolism in the rabbit,
assessed from the blood sugar content, has been investigated in a
wide frequency range — from 9.5 to 9500 kHz (Budko, 1964a).
Only the head of the animal, or the liver region, was exposed in the
field (15 V/cm) of a capacitor for 20 min. Figure 45, which gives
the values of the blood sugar 20 min after the exposure, shows that:
1) Exposure of the head caused much greater changes than exposure of
the liver region; 2) at frequencies of tens to hundreds of kHz the sugar

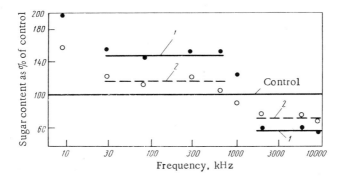

Fig. 45. Change in blood sugar of rabbit due to 20-min exposure
to EmFs of frequency 9.5 kHz-0.5 MHz and strength of 15 V/cm.
1) Irradiation of head; 2) irradiation of liver region.

level was increased, whereas at higher frequencies it was reduced, and
3) within these frequency ranges the magnitude of the effect was
practically independent of the frequency. In the case of all frequencies the sugar level began to return gradually to normal (100%)
20-30 min after the exposure and reached the normal level in 60-
90 min.

However, when the liver tissue was directly exposed (in the
opened abdominal cavity) to the EmF there was no difference in
the effects due to these two frequency ranges; at 9.5 and 9500 kHz
there was an equal and very substantial increase (by a factor of 4.5)
in the sugar level. Subsequent exposures (at hourly intervals) to
a frequency of 9.5 kHz led to an even greater increase in the sugar
level (by a factor of 5.5 after the fourth exposure), but in the case
of 9500 kHz the effect gradually diminished (after the fourth exposure the sugar level was only twice as high as normal).

It is probable that the opposite nature of the effects of low-
frequency and high-frequency EmFs is due to the difference in the
depth of penetration of the fields into the liver and brain tissues
(see Section 4.3).

Multiple exposures of rabbits to 10-centimeter waves of high
intensities for 5-15 min per day upset the regulation of carbohydrate metabolism in the skeletal muscles by reducing the synthesis
of glycogen. The authors (Dainatto et al., 1962) attributed this effect to observed changes in enzymatic processes (particularly the
processes catalyzed by adenosine phosphatase and adenosine di-
phosphatase). In experiments with rats (Kirchev et al., 1962) similar changes, accompanied by a considerable increase (by 68%) in
phosphorylase activity, were observed.

An effect of SHF fields on oxidation — reduction processes in
rabbit organs (liver, kidneys, heart muscle, skeletal muscles, and
brain) has been observed after exposure to high and low intensities
(Moskalyuk, 1957). Exposure of the animals to intensities of 100-
300 mW/cm^2 led to a pronounced reduction in oxidation−reduc-
tion processes, whereas exposure to an intensity of 5-10 mW/cm^2
led to an increase. Multiple exposures caused similar changes,
but they were less pronounced.

A disturbance of carbohydrate metabolism in human beings
due to chronic exposure to low-intensity SHF fields has also been

reported (Bartonicek and Klimkova, 1964). The sugar content of the blood and urine was increased in 75% of the subjects and the sugar curves had a prediabetic form.

An effect of SHF fields on enzyme activity was detected in specially conducted experiments. For instance, exposure of guinea pigs to a relatively high intensity for 5-20 min led to a significant reduction in amylase and lipase activity, as well to a two-phase change — an initial increase followed by a decrease — in the amount of glutathione — an activator of several enzymes (Sacchitelli, 1956, 1958).

The effect of EmFs in the low-frequency and UHF ranges on enzyme activity was investigated (Chirkov, 1964, 1965) in experiments with rabbits. In the first series of experiments the animals were subjected to a single 20-min exposure to frequencies of 8 kHz and 27 MHz at field strengths of 1-20 V/cm. There was a reduction in blood catalase and peroxidase activity only in the case of 9.5 and 27 MHz (the samples were taken at 5-min intervals over a period of 4 h). The reduction of catalase activity due to a field strength of 7-10 V/cm was 8.6 and 9.7%, respectively, in comparison with the control (P < 0.01). The second series of experiments involved multiple exposures to the same frequencies and a field strength of 20 V/cm for 20 min per day (a total of ten exposures at two-day intervals). There was a two-phase change in enzyme activity — a reduction after the first exposure and an increase after the subsequent ones. At frequency 9.5 MHz the catalase activity was altered by -11.5 and +12.6% (after the eighth exposure) and the peroxidase activity by -12.7 and 25.7%. The enzyme activity returned to normal within 2-12 days after the end of irradiation.

It is known that the blood composition and carbohydrate metabolism are regulated in the organism by the hormonal activity of the adrenal cortex. In view of this the investigations of the effect of EmFs on the functions of this gland are of interest. In experiments with 10-centimeter waves (Leites and and Skurikhina, 1961) rats were exposed for 10 min to an intensity of about 100 mW/cm^2. The animals were then killed at different times (1 h to 14 days) after irradiation and the ascorbic acid and lipid content (which provides an index of hormonal activity) of the adrenal cortex was determined. It was found that during the first day after irradiation the amount of these substances was reduced to 70% of the normal level, on the next day it rose to the normal level and subsequently exceeded the normal level by 6-7%. There was a return to normal in two weeks. Similar changes were observed in rats after a 5-min exposure to 3-centimeter waves with an intensity of 400 mW/cm^2 (Gorodetskaya, 1961). We should mention here that the functional activity of the thyroid gland is enhanced to some extent in people who have been exposed chronically to low-intensity SHF fields (Smirnova and Sadchikova, 1960).

An effect on the immune properties of the animal organism has been detected in experiments with a constant magnetic field and low-frequency EmFs.

In mice the amount of antibody protein produced in response to an injection of sheep erythrocytes was reduced by exposure to a constant field of 4000 Oe (Gross, 1962, 1963). Further investigations of this effect with a field of 7000 Oe (Shternberg, 1966) showed that it was most pronounced when exposure to the field occurred at the same time as immunization or the day after it.

Suppression of immunity and production of antibodies to the virus of tick encephalitis has been observed in mice, rats, and rabbits after exposure to a constant magnetic field (7000 Oe) or an EmF with a frequency of 50 Hz and a strength of 200 Oe (Vasil'ev, 1965). However, there was no such effect in the case of production of antibodies to a corpuscular antigen (heterogeneous erythrocytes) and in several cases antibody production was even stimulated. Chronic exposure of white mice (for 14 days) to a field of 2000 Oe slightly inhibited the production of antibodies to typhoid antigens.

In another series of investigations Odintsov (1965) determined the immunobiological indices in experiments in which microbes (Listeria) were injected into mice and guinea pigs which had been exposed once (for 6.5 h) and chronically (for 15 days) to an EmF of frequency 50 Hz and strength 200 Oe. The single exposure did not affect the lethal dose of microbes, their spread through the organism, the number of leucocytes, or their phagocytic activity, but the multiple exposures reduced the natural resistance of the animal organism to Listeria; the phagocytic activity and total number of leucocytes were reduced.

Lantsman (1965) carried out a series of experiments to determine the effect of low-frequency (50 Hz) EmFs on the phagocytic

activity of the reticuloendothelial system, which plays an important
role in the defensive functions of organisms. Mice were exposed to
field of 200 Oe once (for 7 h) or chronically (8 h daily for 4 days).

The single exposure stimulated phagocytic activity, whereas
the multiple exposures inhibited it.

Thus, the effects of EmFs on humoral regulation are characterized by the same common features as all the other biological effects of EmFs — cumulative action, nature of effect practically independent of the EmF frequency, and two-phase nature of the observed changes dependent on the EmF intensity. We can postulate
that most of the discussed disturbances of humoral regulation are
due to reflex or direct action of the EmF on the central nervous
system. Yet some effects indicate that EmFs have a direct effect
on cell function in organs implicated in humoral regulation and also on the biochemical processes responsible for such regulation.
As we will show later, effects of such kind have been directly detected in experiments in vitro.

In concluding this chapter we will point out that almost all
the described effects of EmFs on neurohumoral regulation in entire
organisms clearly show the following common features:

1. The changes in the organism due to EmFs are nonspecific:
They are the same disturbances of neurohumoral regulation which
are produced by a very wide range of other factors.

2. These changes are due mainly to the effect of EmFs on
different divisions of the nervous system. The direct action of
EmFs on the central divisions usually leads to inhibitory reactions,
while direct action on the peripheral divisions usually leads to excitation reactions. The physiological processes controlled by the
nervous system are altered in a corresponding manner.

3. When an EmF acts on a particular division of the nervous
system (which depends on the region of body exposed and the depth
of penetration of the EmF of the particular frequency range) the
nature of the changes produced in the organism is practically independent of the frequency.

4. The nature and magnitude of the changes due to exposure
of the peripheral divisions of the nervous system to EmFs are
practically independent of the EmF intensity. In the case of expo-

sure of the central divisions, however, the effects depend greatly on the intensity. It is characteristic that the central nervous system reacts more strongly to low intensities than to high intensities and in some cases a reaction is observed only at certain low intensities and is totally absent at higher intensities.

5. When an EmF acts both on the central and peripheral divisions of the nervous system there are certain "optimum" intensities (usually two) at which the reaction of the organism is greatest, whereas at other intensities it may be absent altogether. The strength of the reaction varies in a similar manner with the duration of the exposure.

6. The reactions to exposure in the same conditions have two phases, which depend on the intensity and duration of exposure: At low intensities (or short exposures) the changes in the organism are the reverse of those produced at high intensities (or long exposures).

7. The effects of multiple exposures on the organism are cumulative. In this case exposure to strong fields usually leads to adaptation to subsequent exposures, whereas exposure to weak fields leads to progressively greater changes in the organism, which often have a two-phase nature, depending on the number of exposures.

Chapter 8

EFFECT OF EmFs ON REPRODUCTION
AND DEVELOPMENT OF ORGANISMS

In previous chapters we considered the various manifestations of the action of EmFs on normal, fully-formed organisms with completely developed mechanisms of adaptation to the external medium and defense against uncongenial factors. It is natural to expect that the effect of EmFs on the formative processes of the organism — in the embryonic cell, in the developing embryo, and in the growing organism — will be even more significant. Effects of this kind are discussed in this chapter.

8.1. Genetic Effects

The effect of EmFs on the genetic apparatus was first discovered in experiments with growing garlic roots (Heller and Teixeria-Pinto, 1959). The roots were exposed to pulsed UHF fields
in the following conditions: frequency range, 5-40 MHz, pulse
length, $15-50\mu$sec; pulse repetition rate, 500-1000 pulses/sec; field
strength (in pulse), 250-6000 V/m; and duration of exposure, 5 min.
Such treatment led to chromosome aberrations in the garlic root
cells — the formation of bridges and fragments and the formation of
micronuclei. Photomicrographs illustrating these effects are shown
in Fig. 46.

In experiments on Drosophila (Mickey, 1963) similar treatment caused effects of two types: pathological somatic changes,
not hereditarily transmitted, and hereditarily transmitted changes
in the germ cells. An example of effects of the first type was the
formation of red or brown spots on one or both eyes. Effects of
the second type, observed when irradiated adult males were mated
with virgin females, were very diverse: Various sex-linked recessive lethal mutations were almost 13 times more frequent than
in the controls. There was a much higher frequency of visible sex-
linked mutations, such as corneous deformation of the eyes, vesiculation of the wings, reduction in the size of the setae, yellow color
of the body, and dominant changes in the chromosome apparatus.
All these mutations (apart from yellow color of the body) occurred
in several cases, which indicates that some gene loci are particularly sensitive to EmFs. Wing vesiculation occurred both as a

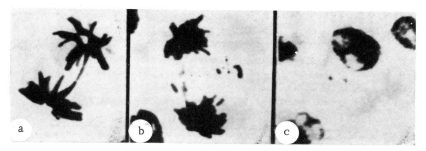

Fig. 46. Changes in chromosome apparatus of dividing cells of garlic root due to
a pulsed UHF field. a) Bridges between daughter groups of chromosomes; b) bridges
and fragments; c) micronuclei formed from fragments.

dominant and as a recessive mutation. The frequency of dominant mutations (due to general aberration of the chromosomes or loss of small parts and duplication) was much higher than in the control. These included changes in the size of the setae, fecundity, reproductive ability, and so on. Mosaics (individuals with altered characters in only some parts of the body), which result from mutations in cells of certain types, occurred.

A further series of investigations by the same authors (at the New England Institute for Medical Research) showed that mutagenic effects similar to those of ionizing radiations could be produced in bacteria, spores of lower fungi and higher plants, insects, and other animals by exposure to pulsed EmFs with particular parameters (Heller, 1963):

1. Exposure of malignant tumor cells to an EmF with a frequency of 30 MHz gives rise to chromosome aberrations, which lead to the cessation of mitosis, aging, and death of the cells. Other frequencies can produce the whole spectrum of chromsome aberrations.

2. Exposure of bacteria which are incapable of synthesizing histidine to an EmF of frequency 31 MHz gives rise to mutants of the opposite type. Bacteria which ferment lactose lose this ability after exposure to a field of frequency 18 MHz. This effect persists in subsequent generations, but when later generations are exposed to 22-MHz waves the fermentative ability is restored and is genetically transmitted.

3. Two kinds of spores of the ascomycete Penicillium, one of which is a mutant *(Penicillium chrysogenium)*, can be separated by their orientation in an EmF. This effect is produced by EmFs of frequencies 11 and 2 MHz and various threshold strengths.

4. When sperm or egg cells of sexually mature Drosophila are exposed in vivo to EmFs with a frequency of about 25 MHz the first generation contains eight times as many females as males. Irradiation with 30-MHz waves leads to changes in the second generation — there are twice as many males as females. On exposure to 28-MHz waves the two dominant genes which determine the eye color of the mated animals produce an unusual recessive.

Each of the described mutational effects is produced by EmFs with particular combinations of parameters: These parameters lie within the following ranges: frequency 1-250 MHz; pulse length, 1-10 μsec; pulse repetition rate, 30-10 000 pulses/sec; and field strength (in pulse), from hundreds to tens of thousands of V/m. The pulse parameters are chosen so that no significant heating of the object occurs and the frequency and strength are chosen in accordance with the desired effect. The effect is produced at a distance, without direct contact with the object.

Genetic effects due to a constant magnetic field have also been observed. Most of the investigations have been made on Drosophila.

Exposure of Drosophila to a field of 3000-4400 Oe (Mulay, 1964) led to a much larger number of cases of corneous deformation of the eye in the first generation (85% against $2 \cdot 10^{-4}$ in the control), but this characteristic was not hereditarily transmitted. Exposure to a magnetic field of 750-1100 Oe (Akhmerov et al., 1966) led to inheritable changes in the form of an increased yield of pupae and flies (by 36% in comparison with the control) in the first and second generations. In other experiments Shakhbazov et al. (1966) investigated the fecundity and viability of the progeny produced by mating fruit flies which had been exposed to a magnetic field with control individuals. Exposure of inbred animals to a field of a field of 1900 Oe increased the fecundity and viability in the second generation by 15% in comparison with the values observed when untreated pairs were mated. Similar changes were observed in the first generation (8-10% increase in comparison with the control) when the two mated parents had been exposed to a field of the same strength. The opposite effect, however — a reduction in fecundity — was observed in hybrids between strains. This change occurred in the first and second generations after exposure to 7000 Oe.

Investigations with magnetic fields of much greater strength — 140,000 Oe (Beischer, 1964) — did not produce any genetic effects in Drosophila: Exposure of pupae for 6 days, young flies for 8 days, and adults for 20 days did not lead to any appreciable changes. Yet the mutation rate of ascomycetes which had been exposed to a field of equally high strength was much higher than in the control (Knepton and Beischer, 1964).

A mutagenic effect of a magnetic field has also been discovered in experiments on higher plants (Pozolotin, 1965; Pozolotin and Gatiyattulina, 1966). Exposure of pea seeds which had previously been irradiated with γ rays (10,000 r) to a pulsed magnetic field of 200,000 Oe led to a statistically significant increase in the number of cells with chromosome aberrations (fragments). A similar treatment of pea seedlings also increased the number of chromosome aberrations.

The experimental discovery of genetic effects of EmFs is itself unexpected. So far a mutagenic effect of electromagnetic radiation has been observed (and regarded as theoretically possible) only for that region of the electromagnetic spectrum in which the quantum energies are high ($h\nu \gg kT$), i.e., for gamma rays, x rays, and ultraviolet rays. An increase in the mutation rate due to temperature increase (by infrared radiation, in particular) has also been observed. Hence, the genetic effects of high-intensity EmFs might be attributed to heating of the cells. But how are we to explain the genetic effects of weak EmFs and a constant magnetic field? There is as yet no adequate experimental evidence for the settlement of this question.

8.2 Effect on Reproductive
Processes

In Chapter 6 we described the various irreversible morphological changes (usually of a degenerative nature) produced in the testicles by high- and low-intensity EmFs and by a constant mangetic field. Here we will discuss the experimental evidence of functional changes in the reproductive organs due to EmFs and the effect EmFs on the sexual cycles and fecundity of animals.

In experiments on rats Gunn et al. (1961a, 1961b) found that a SHF field (24,000 MHz) affected the functions of the sex glands without producing any appreciable morphological changes in the testicles. In view of a previously established effect, the concentration of Zn^{65} injected into males, in the prostate gland, these authors undertook comparative investigations of the amount of this isotope in the prostate gland of animals which had been exposed to a SHF field, infrared rays, or not treated at all. Figure 47 shows a diagram illustrating the considerable reduction in Zn^{65} content after a 5-min exposure to a SHF field and the absence of such an effect on irradiation with infrared rays, which heated the tissue to the same temperature (41°C). The authors concluded that the increase in temperature per se had no effect on the functions of the sexual system and that the observed effect must be regarded as nonthermal.

In another series of experiments (Ciecura and Minecki, 1964) in which the testicles of rats were exposed to SHF fields in the 3000–MHz range a reduction in the activity of several enzymes was observed. A single exposure to an intensity of 64 or 94 mW/cm^2, or multiple exposures to an intensity of 64 mW/cm^2 for 2 min daily for 6 weeks, led to a reduction in the activity of several enzymes (alkaline phosphatase, acid phosphatase, adenosine triphosphatase, and 5-nucleotidase) in the germinal epithelial cells (which produce the spermatozoa). The activity of these enzymes in interstitial cells (which produce the hormone testosterone) was not affected.

An effect of SHF fields on the estral cycle of female mice, the course of pregnancy, and the development of the progeny has been observed in several investigations.

In one series of experiments (Povzhitkov et al., 1961) mice before mating were subjected to multiple exposures of a pulsed SHF field of intensity 0.3 mW/cm^2 for 30 min per day for 20 or 50 days. In another series pregnant females were exposed for 10 min per day to a field of intensity 50 mW/cm^2 for 12 days after mating. This treatment delayed the birth of the young by one or two days, retarded their development, and even led to the death of some of them in the third week after birth.

In experiments with pulsed SHF fields in the 10,000–MHz range (Gorodetskaya, 1963, 1964b) male and female mice were exposed for 5 min to a field of intensity 400 mW/cm^2. When nonirra-

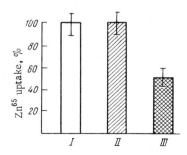

Fig. 47. Comparison of effects of SHF and infrared irradiation on the uptake of Zn^{65} by the prostate gland in rats. I) Not irradiated; II) irradiated with infrared; III) irradiated with SHF.

diated females were mated with the males (just after irradiation)
the percentage of females bearing young was reduced by a factor
of 2.5 and the number of young per litter was reduced by a factor of
1.5 in comparison with the control. Mating on the fifth day after
irradiation led to a much smaller effect and when mating occurred
ten days after irradiation there were no differences in comparison
with the control animals. A similar situation was observed in regard to the number of stillborn mice fathered by irradiated males.
Such effects were much more pronounced when the females were
irradiated: Reduced fecundity, reduced number of progeny in the
litter, and an increase in the number of stillborn young were observed when the females were mated with nonirradiated males just
after, on the 5th day after, or on the 10th day after, irradiation.
Females which had been irradiated showed an increase of approximately 18% in the mean length of the estral cycle (usually due to
prolongation of the rest—diestrus stage).

Investigations of the effect of a magnetic field on mice (Barnothy, 1963a) showed that pregnant females exposed to a field of
2500 Oe produced healthy progeny, but the young mice were all 20%
smaller than those in previous litters of these females. Exposure
to 3100 Oe caused consideralbe disturbances: The newborn mice
died within a few days. When the mice were exposed to 4200 Oe
the embryos were resorbed in the uterus.

The fecundity of insects is affected by an electrostatic or a
magnetostatic field. Interesting results were obtained in experiments in which geometrid moths were exposed for a long period
to an electric field or to an intermittent field switched on and off
every 5 min (Edwards, 1961). The pupae were put into a wooden
box in a field of 180 V/cm between the plates of a capacitor (with
the negative plate grounded); in the control box both plates were
grounded. After a week's exposure to the constant field the emergence of the adults was much later than in the control; in the case
of the intermittent field, however, the emergence was only slightly
later (less than a day) than in the control. A considerable difference
was discovered in the number of eggs laid by females in the presence
and absence of the field: Firstly, in the presence of the field the
total number of eggs laid was less than in the control; secondly,
a much greater number of eggs (per female) was laid on the outer
(positively charged) surface of the capacitor plate than on its inner

surface, where the field was stronger. Thus, an electric field adversely affects the development of pupae into adult moths, and in laying eggs the adult moths try to avoid the field.

A magnetic field has the opposite effect on the reproduction of insects (Kogan et al., 1965). When Drosophila are placed between the poles of a magnet they collect near the poles and lay more eggs there than in the space between the poles.

Thus, SHF fields, and also magnetic and electric fields, have an inhibitive effect on reproductive processes and upset their normal course. Such effects have also been observed in experiments with bacteria, as will be related in the next chapter.

8.3. Effect on Embryonic Development of Vertebrates

Several investigations have shown that SHF fields affect the development of the chick embryo. The first experimental evidence of such an effect was obtained as long ago as 1940 (Van Everdingen). Exposure of eggs containing a 5-day-old embryo to a SHF field with a frequency of 1875 MHz reduced the rate of metabolism by a factor of 1.5 and caused the death of the embryo. Irradiation at later stages of development of the embryo had less effect on the metabolism and did not cause death, and an 11-day old embryo did not show any reaction to irradiation. A SHF field also affects the heart rate of the embryo, beginning at the period when glycogen (associated with carbohydrate metabolism) appears in the heart: The heart rate is reduced from 90–110 to 10–20 beats per min and the amplitude of the ECG spikes becomes greater than normal.

The effect of SHF fields on the heart activity of the embryo has been investigated more thoroughly in recent experiments (Paff et al., 1962, 1963).

The experiments were conducted on 106 embryo heart preparations after 72 h of incubation. The ECG was recorded continuously for the 3 min before irradiation, during irradiation, and 3 min after it. The preparations were exposed to a SHF field of frequency 24000 MHz and intensities of 478, 297, 167, or 74 mW/cm^2 for a few seconds to 3 min. During irradiation the preparations were cooled by a stream of air to prevent their overheating. Exposure

to 167 mW/cm^2, which raised the temperature of the preparation
to 38 deg, caused appreciable disturbances in the ECG: shortening
of the QT interval, an increase in the amplitude and width of the
T wave, and an increase in the U wave. At intensity 74 mW/cm^2
the temperature rose only to 25°C, but the changes in the ECG
were the same.

Several experiments have shown pronounced disturbance of
the development of the embryo due to exposure to SHF fields (Carpenter, 1960; Van Ummersen, 1961). Eggs which had been incubated
for 48 h were exposed to a SHF field of frequency 2450 MHz and an
intensity which caused heating to the same temperature as incubation (39 deg). This led to inhibition of the normal development of
the embryo: In structures which were already differentiated only
proliferation (cell multiplication) took place and there was no further differentiation, and in structures which had not become differentiated proliferation ceased and the embryo stopped developing.

An effect of a weak magnetic field (4-7 Oe) on the development of the pigeon embryo has been discovered (Kiryushkin et al.,
1966). In this bird the egg laid first usually weighs less than the
second, loses more weight during incubation, and hatches earlier.
Exposure to a magnetic field led to the reverse situation: the
loss of weight in the incubation period was greater in the second
egg than in the first, and the second egg hatched earlier than the
first. Toroptsev et al., (1966b) observed an inhibitive effect of a
magnetic field of 7000 Oe on the embryonic development of frogs.

Thus, a magnetic field and SHF fields upset normal embryonic development — they adversely affect metabolic processes and
inhibit cell multiplication and differentiation.

8.4. Effect on Growth and
Development of Organisms

Investigations of the effect of a magnetic field on the development of mice were begun as long ago as 1948 at the Budapest Institute
of Experimental Physics and were pursued further by the Barnothys
(J. Barnothy, 1963; M. Barnothy, 1964; J. Barnothy and M. Barnothy, 1963).

In one series of investigations 5-day-old mice were kept for
four weeks in a vertically directed magnetic field of 5900 Oe (gra-

dient 100 Oe/cm) and their increase in weight during this period
and the following four weeks was recorded. The results are illustrated in Fig. 48. As the curve shows, the weight of the animals
which had been exposed to the magnetic field lagged behind that of
the controls throughout the experiment and did not reach normal
until the end of the experiment. On the second day there was a
pronounced loss of weight, which the authors believed to be due to
a "shock state" induced by the field. Some of the males showed a
marked loss of weight on the 11th day and this sometimes led to
their death. The females did not show any unfavorable signs; mating (with normal males) after exposure to the field led to normal
gestation and the birth of normal young.

In the second series of investigations the development of
young (30-day-old) and adult (60-day-old) animals which had been
exposed to a magnetic field of 4200 Oe with a gradient of 80 Oe/cm
or of 3600 Oe with a gradient of 650 Oe/cm (the graidients were determined relative to the centers of gravity of the mice's bodies) was observed. These experiments, conducted on 680 animals, gave the
following results:

1. In all the exposure conditions the magnetic field led to a
lower rate of increase of weight than in the control.

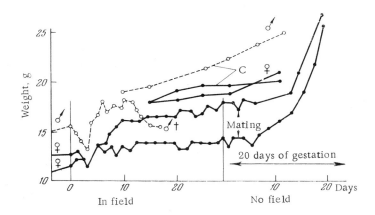

Fig. 48. Effect of a constant magnetic field of
5900 Oe on the growth of 5-day-old mice: ♂)
Males; ♀) Females; C) control.

   2. The field with the low gradient (4200 Oe, 80 Oe/cm) had a greater effect of this kind than the field with the high gradient (3600 Oe, 650 Oe/cm).

   3. The mean difference in the weight of the experimental and control mice was greater in young animals than in the adults, but the relative reduction was more pronounced in adults.

   4. Individual differences were greater in the experimental group than in the control, which indicates that the field acts differently on different individuals.

   5. On the second day of exposure to either field there was a marked loss of weight.

   The last effect — "the second-day minimum" — was specially investigated in 5-week-old animals over a period of 10 weeks. During this period two groups of animals were alternately exposed for four days to a field of 9400 Oe and for days in similar conditions (model of magnet) without the field. The results are summed up in Fig. 49, which shows that the pronounced drop in weight on the second day was distinctly manifested in both groups (P < 0.0001). The magnitude of the minimum did not decrease in the subsequent cycles and, hence, the animals did not become adapted to the "shock" induced by the field.

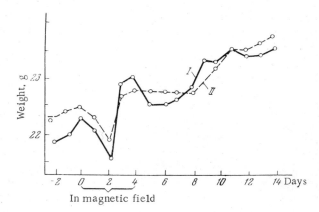

Fig. 49. Illustration of marked loss of weight of mice on second day of exposure to a magnetic field of 9400 Oe. I and II) Groups exposed alternately for 4 days to magnetic field and for 4 days without field.

An effect of multiple exposures to a SHF field on the growth of rats has been discovered (Lobanova, 1960). The animals were exposed to fields of 3000 MHz and intensities of 10, 40, and 100 mW/cm^2 for 60, 15, and 5 min per day (corresponding to the increase in intensity) for four weeks. Figure 50 shows the weight curves (mean data for 45 animals), which illustrate the slightly greater weight of irradiated animals than the controls; this difference increased after the end of irradiation. Thus, SHF fields (in the given exposure conditions), in contrast to a magnetic field, have a slight stimulating effect on the growth of animals.

In connection with the described manifestations of the action of EmFs on the reproduction and development of animals, investigations of the effect of SHF fields on the DNA and RNA content of animal tissues and organs, and of the activity of the corresponding enzymes — ribonuclease (RNAase) and deoxyribonuclease (DNAase) — were undertaken. In experiments with SHF fields of 10,000 MHz (Kerova, 1964) rats were irradiated for 6 min with intensities of 100 and 500 mW/cm^2. It was found that SHF fields reduced the activity of DNAase and RNAase; the RNA content was increased, but the DNA content was reduced. Infrared irradiation caused changes of the same nature but the changes were less pronounced.

Investigations of the effect of SHF fields of nonthermal intensities (3000 MHz, 10 mW/cm^2) on nucleic acid content were made on rats subjected to chronic exposure (Nikogosyan, 1964c). After 40-60 daily exposures for 1 h there was a reduction in the nucleic acid content only in the spleen, while after 80 exposures there was a reduction in the brain and liver too. After 120-140 exposures the RNA content began to increase again and 30-35 days after the end of irradiation it was completely normal. In all the experiments the DNA content was unaltered.

The effects of a magnetic field on the development of plants have been studied for more than 60 years. As far back as 1903 Ewart found that if aquatic plants (Vallisneria and Chara) were placed in a magnetic field so that the streaming of the protoplasm was perpendicular to the lines of force, the streaming become slower or even stopped; there was no effect in the case of parallel motion. Later, Savostin (1928, 1937) confirmed the effect of a perpendicular field, but he also observed a reduction of 15-30% in the rate of streaming of the protoplasm in a strong parallel field

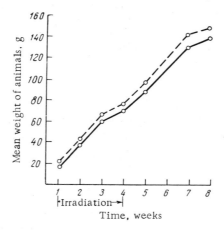

Fig. 50. Effect of chronic exposure to a SHF field on increase in weight of rats. The continuous curve is the control and the broken curve represents the irradiated animals.

(7000 Oe). He found that in a perpendicular field acceleration of the streaming of the protoplasm was more frequent than retardation.

Savostin also found that a magnetic field increased the rate of growth of plant radicles and the permeability of the cell walls. Later investigations confirmed these effects and also revealed other effects of a magnetic field on plants.

It has been found that a magnetic field increases the yield of tomatoes (Karmilov, 1948) and accelerates their ripening (Boe and Salunkhe, 1963), and that exposure to a field of only 20–60 Oe increases the growth of the root system of rye and beans, and leads to more rapid germination of wheat and corn seeds (Krylov and Tarakanova, 1960; Pittman, 1963). Experiments with barley (Mericle et al., 1964) provide an illustration of these effects: In a magnetic field of 1200 Oe the roots and shoots of barley seedlings grow more rapidly than normal (Fig. 51).

Attempts to discover the biochemical mechanism of the effect of weak and strong magnetic fields on plants (Novitskii et al., 1965, 1966; Novitskii, 1966a, 1966b) led to the following results:

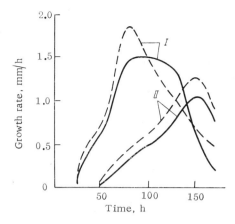

Fig. 51. Effect of magnetic field of 1200
Oe on growth rate of roots(I) and shoots
(II) of barley seedlings. The broken curves
are for the magnetic field and the continuous curves are for the control.

1. The stimulating effect of weak magnetic fields (20–60 Oe)
on plant growth is particularly pronounced in the first two or three
days of growth of the seeds — in the period of relatively low enzymatic activity; in this period the action of the field leads to reduced oxygen consumption (to 20–25% in two-day-old rye seedlings),
increases the rate of mitosis in the roots and stems, and increases
the DNA content of plant cells.

2. The action of a field of 80 Oe on an Elodea leaf accelerates
the movement of the chloroplasts in the stream of protoplasm (in
90% of the cases in summer and in 52% in autumn); the electrical
resistance of gels of substances of plant origin is reduced — by
5–15% in the case of agar–agar and by 5–40% in the case of starch.

3. Exposure to a strong field (4500 Oe, 1 h) parallel to the
axis of cereal seeds reduces their magnetic susceptibility by 20%
in comparison with control seeds (oriented in the direction of the
earth's magnetic field).

Other investigations (Tarchevskii, 1964; Zabotin, 1965;
Zabotin and Neustroeva, 1966) showed a reduced rate of photosynthesis in Elodea and wheat leaves.

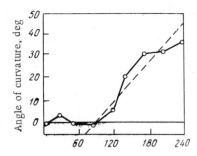

Fig. 52. Curvature of root of cress growing in a magnetic field (towards lower
field strength) in relation to time of exposure.

The phenomenon of "magnetotropism" — the oriented growth
of the root system in a magnetic field — has been observed. For
instance, corn radicles on germination bend towards the south magnetic pole (Krylov and Tarakanova, 1960), while rootlets of cress
and corn, oat coleoptiles, and sunflower hypocotyls bend in the direction of lower strength of the magnetic field (Audus, 1960), as is
illustrated for cress by the plot of the angle of curvature shown in
Fig. 52 (Audus and Whish, 1964).

Experiments with oat seedlings (Pickett and Schrank, 1965)
showed an effect of a magnetic field on the bending of coleoptiles.
In a magnetic field of 565 and 1200 G the bend of the seedling coleoptiles (grown in a clinostat rotating at 1 rpm) was 10.44 deg, whereas it was 15.15 deg in the control, but an electric field of 268-1230
V/cm completely removed the bend. The combined action of a
magnetic and electric field had a lesser effect.

It has recently been found that an electrostatic field also affects plant growth (Murr, 1965). Experiments with sorghum and
orchard grass showed that an electric field inhibited the development of these plants; a high percentage of injuries to the leaf epi-
dermis was also observed. Figure 53 shows a plot of the percentage of injured plants against the "static" field strength (constant
during the growth of the plant). The author points out, however,
that a more strongly acting factor is the "dynamic" field strength
(varying as the plant grows), which gives rise to a negative charge
on the upper leaves of the plant (in the absence of a field all the
leaves have a positive charge). Ion radicals and polarized molecules appeared in the damaged leaves.

The experimental information on the effect of EmFs on the
growth of animals and plants does not allow us to draw any conclu-

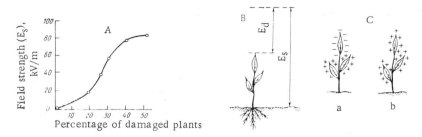

Fig. 53. Effect of electrostatic field on development of sorghum. A) Injury to plants in relation to field strength; B) static (E_s) and dynamic (E_d) field strengths; C) electric charges on leaves in electrostatic field (a) and in absence of field (b).

sions as yet. We can merely point out general features of the action of a magnetic field: It has an inhibiting effect on the growth of animals and a stimulating effect on the growth of plants.

As regards the effect of EmFs on various stages of development of organisms — from the germ cell to the growing organism — we can point to only one general feature: In most experiments EmFs had an adverse effect on these processes.

Chapter 9

EFFECTS OF EmFs AT THE CELLULAR AND MOLECULAR LEVELS

In discussing the manifestations of the action of EmFs on fully-formed entire organisms and on development from the cell to the fully-formed organism we have encountered various effects at the cellular and molecular levels. This naturally gives rise to the questions: What effects can EmFs have on cells and macromolecules outside the organism? How far do these effects differ from those observed at the same levels in entire organisms?

A considerable amount of information relating to these questions has now been gathered (particularly in recent years): The effect of EmFs on isolated tissues and cells, on cell cultures and unicellular organisms, on protein solutions and crystalline proteins, and, finally, the effect of a magnetic field on the physicochemical and biochemical properties of water, have been investigated.

9.1. Effect on Isolated
Tissues and Cells

The effect of EmFs on carbohydrate metabolism in the liver was investigated in experiments with isolated liver (Budko, 1964a). The experiments were conducted in the frequency range 0.5-21,500 kHz at a field strength of 15 V/cm. It was found that low- and high-frequency EmFs reduced the glucose content in comparison with the control, whereas UHF fields increased it. Fig. 54 shows a plot of the effect against frequency (from the mean values of 150 experiments, $P < 0.02$). The variation of the effect with the EmF strength was also investigated and is illustrated graphically in Fig. 55. Thus, the alteration of the carbohydrate metabolism in isolated liver by an EmF depends considerably on the frequency of the EmF and its strength. It is characteristic that the effect does not increase with increasing strength, but attains a maximum at some "optimum" strength.

Since the nervous system of animals is particularly sensitive to EmFs investigations of the direct action of EmFs on isolated nerve cells are of interest. As long ago as 1900 Danilevskii detected the excitation of a frog nerve situated at a distance of several meters from an EmF source (a spark gap), but there have been very few further investigations in this direction.

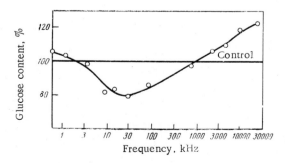

Fig. 54. Change in glucose content of isolated rat liver due to low- and high-frequency EmFs.

Petrov in 1935 observed the excitation or change in excitability of a nerve-muscle preparation due to a low-frequency EmF. However, it was not until recent years that some additiona information about this effect was obtained: It was found that the change in excitability persisted for 1-5 min after stimulation of the nerve by a low-frequency EmF and was of a two-phase nature (Sazonova, 1960) and also that a low-frequency EmF sensitized the nerve to the blocking effect of novocaine and solutions with a high potassium content (Pudovkin, 1964).

More definite evidence of an increase in excitability of a frog nerve—muscle preparation due to SHF fields has been obtained. The first qualitative description of this effect appeared in 1960 (Bychkov and Moreva) and quantitative investigations have recently been carried out (Kamenskii, 1964, 1967). The methods and results of these investigations merit a fuller account.

The preparations were exposed to continuous SHF fields of frequency 2400 MHz at an irradiation intensity of 10-1000 mW/cm^2 and pulsed SHF fields of frequency 3000 MHz with a pulse length of 1 μsec, a repetition rate of 100-700 pulses/sec, and mean intensity of about 10 mW/cm^2. Stimulation of the nerve by rectangular dc pulses 0.1 to 1 μsec long twice a second led to a change in five parameters of the functional state of the nerve — the excitation threshold, the amplitude of the biopotentials, the speed of conduction of excitation along the nerve, and the lengths of the absolute and relative refractory phases. The heating of the nerve due to the SHF field was measured at the same time.

These investigations produced the following results:

1. Continuous SHF fields which heated the nerve by 2 deg in 30 min led only to an increase in the speed of conduction of excita-

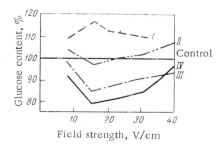

Fig. 55.   Change in glucose content of isolated rat liver in relation to EmF strength. I) 9.5 MHz; II) 730 kHz; III) 85 kHz; IV) 29 kHz.

tion (by $16 \pm 4.5\%$), and a slight shortening of the absolute and relative refractory phases. At higher intensities, when the nerve temperature rose 3-9 deg in 1 min, the increase in speed of conduction was much greater.

It is obvious from a comparison of the graphs (Fig. 56) showing the alteration of this parameter of the nerve by SHF irradiation and ordinary heating that the effect of the SHF field is not a thermal one, but is of some other nature. In addition, exposure to high-intensity SHF fields led to a two-phase change in the amplitude of the biopotentials, as shown in Fig. 56B, and this effect was of a nonthermal nature.

2. Experiments with pulsed SHF fields which heated the nerve by 2 deg in 30 min led to an increase in excitability (lowering of excitation threshold), as shown by the graph in Fig. 56C. There was also a 10% increase in the speed of conduction of excitation.

An effect of a constant magnetic field on nerve and muscle cells has also been detected. Exposure of the myocardial fibrils of a common snail (in a moist chamber) to a field of 15,000 Oe led to a reduction of diastolic activity and the heart rate; when the field was removed there was a short-lived increase in these values above the initial level (Chalazonits and Arvanitaki, 1965). In investigations with a frog nerve-muscle preparation (Aminaev and Khasanova, 1966) the gastrocnemius, exhausted by rhythmic electric stimulation of the nerve at a rate of about 2 pulses/sec, was exposed to a field of 300-800 Oe. In these conditions exhaustion of the muscle and the onset of myoneural shock occurred later than in the control.

The effect of a magnetic field (2000 Oe) on the spontaneous bioelectric activity of the isolated nerve cord of the freshwater crayfish has been observed (Luk'yanova, 1966). Experiments on 50 preparations showed that there was an increase in activity in 70% cases in the autumn period ($P < 0.05$), and a reduction in 76% cases in winter ($P < 0.01$).

9.2. Effect on Cell Cultures

The effect of EmFs on cell cultures can either be stimulating or inhibiting, depending on the kind of culture and the EmF frequency.

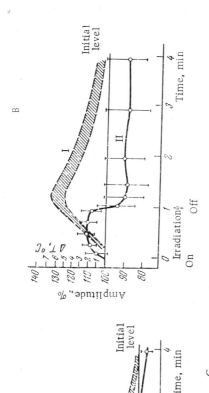

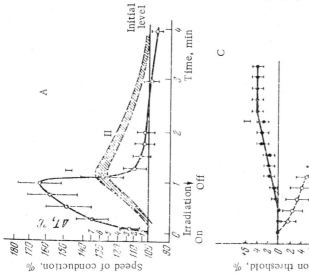

Fig. 56. Changes in speed of conduction of excitation (A), amplitude of biopotentials (B), and excitation threshold (C) of frog nerve due to SHF fields. I) Exposure to SHF; II) control.

A stimulating effect of SHF fields was found in experiments with a chick-embryo heart-tissue culture (Seguin et al., 1948, 1949). Exposure to low-intenisty SHF fields (1400 and 1000 MHz), which caused slight heating of the culture, led to enhanced growth in comparison with the control, which was heated to the same temperature by infrared radiation.

An inhibiting effect of low-frequency EmFs was found in experiments with cultures of normal and malignant human cells (Knoepp et al., 1962). The cultures were exposed to EmFs in the frequency range 99-1000 Hz at a field strength of 1.1-1.7 V/m for 1-3 h. This treatment raised the temperature of the cultures by only 2.3 deg, but their growth was inhibited — from a very slight retardation to complete cessation of the growth of the culture and death of the cells. The fact that these effects occurred only at particular frequencies, specific for each type of cell, is of particular interest.

Several investigations have shown an inhibiting effect of a magnetic field on cultures of malignant cells. Exposure of a culture of human nasopharyngeal cancer cells (KB cells) to a magnetic field of 4000 Oe led to a reduction of $9 \pm 7\%$ in the number of cells on the third day, as compared with a $31 \pm 11\%$ increase in the number of cells in the control culture (Butler and Dean, 1964). Cells of a sarcoma 37 ascites culture degenerated after an 18-h exposure to a magnetic field of 4000-8000 Oe at a temperature of 37°C (Mulay and Mulay, 1961; Mulay, 1964). Exposure of a culture of Ehrlich ascites tumor cells to a magnetic field of 7300 Oe for 1-3 h (at a temperature of 37°C) greatly reduced (to 52%) the rate of oxygen exchange (Reno and Nutini, 1963).

Yet the inhibiting effects of UHF and SHF fields on malignant tumors in vivo, as described in Section 6.4, have not been found in the corresponding cell cultures: Exposure of an Ehrlich ascites tumor culture to UHF fields (Merli et al., 1963) and of a mouse sarcoma 180 culture to SHF fields (Moressi, 1964) led only to the thermal effect.

An investigation of the effect of a magnetic field on human blood in vitro (Mogendovich and Tishankin, 1948a, 1948b; Mogendovich and Sherstneva, 1947, 1948a, 1948b; Mogendovich, 1965) revealed the following effects:

1. The erythrocyte sedimentation rate (ESR) was reduced and a rotational motion was imparted to the erythrocytes.

2. The rate of descent of a drop of blood in a copper sulfate solution (of equal density) was increased owing to a change in the permeability of surface of the drop.

3. The clotting time was increased.

An effect of a magnetic field on the specific agglutination of human erythrocytes was recently discovered (Hackel et al., 1961, 1964; Foner, 1963; Hackel, 1964). The effect of magnetic fields of 23 to 18,000 G on the percentage agglutination of erythrocytes of a particular blood group in the presence of sera (agglutinins) which usually agglutinate these erythrocytes, was investigated. An increase in the percentage agglutination in comparison with the control was observed at field strengths of more than 53 G; this increase was greatest at 5000-8000 G and became slightly less at higher strengths.

We have already mentioned the orientation of erythrocytes and leucocytes along the electric lines of force of a UHF field (Herrick, 1958; Wildervank et al., 1959). A similar effect due to a magnetic field has recently been detected (Murayama, 1965): In a field of 3500 G sickled erythrocytes (pathologically altered erythrocytes characteristic of sickle-cell anemia) oriented themselves perpendicular to the magnetic lines of force.

9.3. Effect on Unicellular Organisms

The effects of EmFs on unicellular organisms can be divided into two groups: 1) a particular orientation or directed motion of unicellular organisms due, presumably, to the action of the EmFs on their "peripheral excitable structure"; 2) a change in physiological functions, which might be due to the action of the EmF on the "central" (intracellular) control systems of unicellular organisms.

The oriented motion of unicellular organisms in the direction of the lines of force of an electric field and low-frequency EmFs ("electrotaxis" and "oscillotaxis") was discovered at the end of last century (see the review by Scheminzky and Bukatsch, 1941).

However, the main features of this effect were not established until recent years, when the behavior of unicellular organisms in
UHF fields was investigated (Herrick, 1958; Wildervank et al.,
1959; Heller, 1959; Teixeria-Pinto et al., 1960; Heller and Mickey,
1960; Mickey, 1963):

1. Motile unicellular organisms (flagellates, ciliates) orient
themselves and move in a UHF field either parallel to or perpendicular to the electric lines of force, depending on the frequency;
each species of organism responds in its own way to different frequencies. Movement perpendicular to the field usually occurs at
higher frequencies than movement parallel to the field. For instance,
Euglena moves along the lines of force in a field of 6-7 MHz (Fig.
57), and perpendicular to the lines of force in a field of 27-30 MHz.

2. Amebae extend their bodies along the lines of a field with
a frequency of about 5 MHz, but perpendicular to them at a frequency of 27 MHz. Their intracellular asymmetrical particles may
become oriented in a direction prepO endicular to the orientation of
the body in the UHF field. A similar mutually perpendicular orientation of the body and intracellular particles has been observed
in paramecia too.

It was suggested, by comparison with the above-described
(Section 4.5) orientation of particles, that these reactions of unicellular organisms to UHF fields are passive. Further investigations, however, showed that the behavior of unicellular organisms
in a field of any frequency is due to physiological reactions of these
organisms, rather than to purely physical processes. It was found
as long ago as 1936 (Kinosita) that when ciliate infusoria are exposed to a constant electric field the direction of beating of the cilia
on the side of the body facing the cathode is reversed. Later (1954),
the same author found a potential difference between the inner and
outer surface of the body of infusoria (as in a nerve cell), which
periodically altered in time with the beating of the cilia. It was recently shown (Dryl, 1965) that the turning and direction of motion
of ciliate infusoria relative to the lines of force of an electric field
are due to depolarization of the surface membrane, which causes
reversal of the beating of the cilia. The same author found that
the direction of oriented movement of unicellular organisms in a
UHF field depends not only on the frequency, but also on the strength,
of the field. For instance, at a frequency of 11.5 MHz and a field

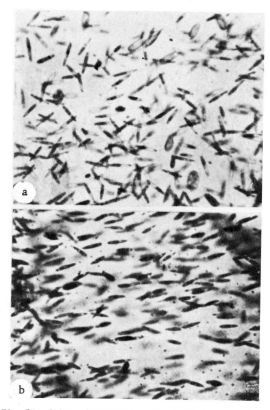

Fig. 57. Oriented movement of unicellular organisms (Euglena) due to EmF of frequency 5-7 MHz. a) Random motion in absence of EmF; b) oriented motion parallel to electric lines of force of EmF.

strength of about 1000 V/cm the unicellular organism *Rhabdomonas incurva* moves along the field lines, Astasia moves perpendicular to the lines, and Colpidium at random. At 27 MHz and a field strength of about 600 V/cm all these unicellular organisms move perpendicular to the field lines.

A series of investigations with paramecia (Presman, 1963b; Presman and Rappeport, 1964a, 1964b; Zubkova, 1967b) provided evidence that these infusoria have an excitable structure which functions like the neuromuscular system of vertebrates. It was found that the infusoria responded to pulses of direct and alternating current of particular threshold strengths by an "electric shock reaction," consisting in a sudden stoppage of motion and turning of the body axis parallel to the electric lines of force. It was found that the relationship between the threshold strength (for the electric shock reaction) and the length of the direct-current pulse, or the frequency of the alternating current, was of the same nature as in the case of similar stimulation of nerve and muscle tissues of vertebrate animals (Fig. 58).

Continuous exposure of paramecia to alternating current of various frequencies and increasing voltage led successively to three kinds of movement of the paramecia and, finally, to their death. These data are given in Table 14, which shows that the reaction of the infusoria increased with frequency increase. This was revealed by the fact that the same cycle of changes in behavior of the paramecia at 5000 and 50,000 Hz required a much smaller range of voltage change (relative to the threshold for the electric shock reaction) than at 50 and 500 Hz. It is of interest that on exposure to an EmF the paramecia perform the same kinds of motion as has been observed in natural conditions (Parducz, 1954).

The results of these investigations (carried out in a wide frequency range — from 20 Hz to 10 MHz) confirmed the earlier hypothesis that paramecia have an excitable structure which is sensitive to EmFs and indicated that paramecia would react in a similar manner to SHF fields, too. Such sensitivity has been discovered (Presman, 1963c; Presman and Rappeport, 1965; Zubkova, 1967a).

It was found that the electric shock reaction could be produced in paramecia by single SHF pulses (2400–3000 MHz) of various lengths or by series of short pulses (1 μsec). The threshold values of the power in the pulse (or the mean power in the series of pulses) at which the reaction occurred were inversely proportional to the length of the pulse (or the length of the series of pulses), as Fig. 59A shows. It was also found that although SHF pulses did not cause the electric shock reaction at subthreshold powers, such pulses increased the sensitivity of paramecia to other stimuli. Such sensitization to stimuli by ac pulses is shown in Fig. 59B.

We will mention the typical features of the considered reactions: 1) The nature of the electric shock motor reaction was the

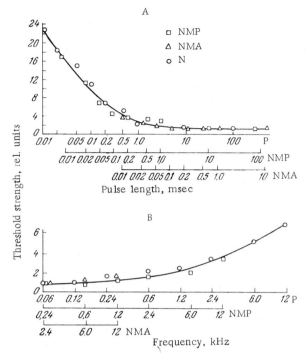

Fig. 58. Comparison of variation of excitation threshold of paramecia (P), frog nerve—muscle preparation (NMP), and human
nerve—muscle apparatus (NMA) with length of the stimulaing
pulses of direct current (A) and frequency of alternating current (B).

same in the case of exposure to alternating currents in wide frequency range and to SHF fields (3000 MHz); 2) the threshold values
of the voltage and power required to produce the electric shock reaction were inversely proportional to the square root of the pulse
length or the frequency (and not to the first powers of these quantities); 3) the heating of the medium containing the paramecia was
insignificant in all the exposures which cause the electric shock
reaction. Thus, the electric shock reaction can be regarded as a
frequency-independent nonthermal effect. Here we again encounter
the same basic features of the biological action of EmFs as were
observed in the reactions of the nervous system of animals to EmFs.

Migration of paramecia in a magnetic field has also been discovered (Kogan and Tikhonova, 1965). Paramecia in a capillary

TABLE 14. Behavior of Paramecia on Continuous Exposure
to EmFs of Different Frequencies in Relation to Voltage

Frequency, Hz	Voltages at which different kinds of behavior are observed							
	Weak rotation of same infusoria in EmF		Rotation, with sudden sharp movements, of all infusoria in EmF		Vibrations of infusoria with body axis perpendicular to electric lines of force		"Polarization" of infusoria, swelling of body and death	
	in volts	relative to electric shock threshold	in volts	relative to electric shock threshold	in volts	relative to electric shock threshold	in volts	relative to electric shock threshold
50	1/1*	2,8/2*	1,7	4,7	4,5/5*	12,5/10*	5,5	15,2
500	1	1,3	1,5	2	6	7,8	8	10,4
5.000	4	1,7	5,5	2,3	—	—	6,7	2,8
50.000	6,2	1	12,5	2	—	—	20	3,3

* The numerator gives the data of the described investigations and the denominator gives the data of Scheminsky et al. (1941).

of diameter 0.5 mm were placed in a field of 700–800 Oe. The infusoria collected at the south pole; 10 min after the field was switched off they became uniformly distributed throughout the capillary again. This accumulation of paramecia at a particular end of the capillary was observed when infusoria were placed in water which had been exposed (in the capillary) to a magnetic field. In other investigations Kogan et al. (1966) found that a magnetic field affected the nature of the motion of the ciliate Stylonychia: its random motion was converted after a regular motion in a circle of radius 600–800 μ; after 3–10 min this motion slowed down considerably and the infusoria came to a stop, turning round their body axis; after 10–20 min they resumed their initial kind of movement, but it was slower.

Effects which might be attributed to the action of EmFs on "central" control systems have been observed in several investigations with bacteria.

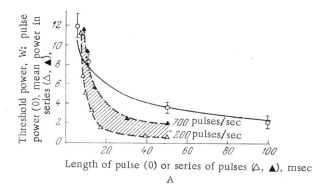

Length of pulse (0) or series of pulses (△, ▲), msec

A

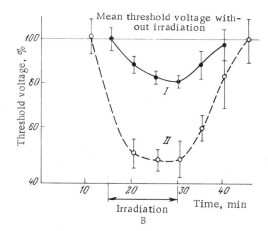

B

Fig. 59. Effect of SHF fields on excitable structure of paramecia.
A) Plot of threshold power of pulses (solid curve) or series of pulses
(hatched area) required to produce electric shock reaction against
length of pulse or series of pulses; B) reduction of threshold voltage
for electric shock reaction on stimulation by ac pulses (1200 Hz, 120
μsec) and simultaneous exposure to SHF pulses: I) Single pulses;
II) series of pulses.

One group of investigations showed bactericidal effects, but
these occurred only at high irradiation intensities, where the heating of the bacterial culture was significant. For instance, when an
Escherichia coli culture was heated to 55-60°C by pulsed fields
in the range from 65 Hz to 600 MHz, the viability of the bacteria
was greatly reduced (Brown and Morrison, 1954, 1956). Irradia-

tion of fowl sarcoma virus at 3000 MHz led to total inactivation, but there was no effect when the culture was simultaneously cooled (Epstein and Cook, 1951). Experiments in which luminous bacteria were exposed to SHF fields in the range 2600–3000 MHz failed to reveal a nonthermal effect (Barber, 1961).

Another group of investigations (more numerous) showed that EmFs affect bacteria in a way which cannot be attributed to the thermal effect. Exposure of cultures of staphylococci, colon bacilli, and Koch's bacilli for 1 min to an EmF of frequency 1400 MHz, which heated the culture to 34°C, arrested the multiplication of the bacteria and the latent period (from the moment of cessation of irradiation) was much shorter than in the case of ordinary heating (Seguin and Castelain, 1947a; Seguin, 1949a). Schastnaya (1955, 1957, 1958) exposed cultures of the colon bacillus, *Staphylococcus aureus*, and Friedlander's bacillus to SHF fields and kept the temperature of the suspension in the range 37–42°C, i.e., below the lethal level for these bacteria. The growth of colonies in the irradiated cultures was poorer than in cultures heated in a water bath to the same temperatures.

Experiments with EmFs of lower frequencies have revealed bactericidal effects: Exposure of colon bacilli to an EmF of frequency 20 MHz and a field strength of 205 V/cm (Nyrop, 1946) led to death within 5–10 sec when the medium was heated to 40°C, whereas simple heating required 10 min at 60 deg to produce this effect. Foot-and-mouth disease viruses were completely inactivated on exposure to a field of 260 V/cm for 10 sec and 480 V/cm for 2.4 sec; the medium in this case was not heated above 37°C. To produce the same effect by simple heating required 60 h. In experiments with EmFs in the 12- to 300 MHz range (Fleming, 1944) a bactericidal effect was manifested in 5 min, although the maximum temperature of the culture did not exceed 37°C.

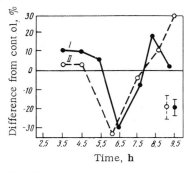

Fig. 60. Change in growth of bacterial culture in nonuniform magnetic field (15,000 Oe, 2300 Oe/cm). I) *Serratia marcescens;* II) *Staphylococcus aureus.*

Interesting observations have been made on the effect of a nonuniform magnetic field of 15,000 Oe with a gradient of 2300 Oe/cm on bacteria (Gerencser et al., 1962, 1964): In these conditions the growth of the cultures relative to the control varied in two phases, the growth was inhibited at first, and then it was enhanced (Fig. 60). A similar, but weaker, effect was found in the case of a uniform magnetic field of 14,000 Oe (Hedrick, 1964).

The recently observed changes in phagocytic activity of para-mecia (determined by the uptake of india-ink particles into the food vacuoles) (Kulin and Morozov, 1964, 1965; Kulin, 1965) due to exposure to a SHF field of frequency 2400 MHz can be attributed to the effects of EmFs on central control systems. As the graphs given in Fig. 61 show, there are four phases of variation of the phagocytic activity, depending on the irradiation intensity — stimu-lation at relatively low intensities, inhibition at moderate intensi-ties, and then the repetition of these phases with further increase in intensity. The same heating of the medium containing the para-mecia by infrared rays or on a water bath led to a two-phase change in phagocytic activity — stimulation at temperatures optimum for these infusoria and inhibition at higher temperatures.

From the described investigations of the effect of EmFs on isolated tissues, cells, and unicellular organisms we can partially answer the questions posed at the beginning of this chapter.

The reactions of unicellular organisms to EmFs show the same general features as those of complex organisms: The nature of the reaction does not depend on the frequency of the EmF; the effect depends on the intensity in cases where it can be assumed that the EmF acts directly on the "central" intracellular control sys-tems and is independent of intensity when the effects are due to the action of the EmF on the peripheral excitable structure of the uni-cellular organisms.

The effects observed in experiments with isolated tissues and cell cultures can be attributed to the direct action of EmFs on the "central" (intracellular) regulation. However, as distinct from the situation observed in the case of irradiation of tissues and cells in the entire organism, the nature and magnitude of the effect in experiments in vitro depend not only on the intensity of the EmF, but also on its frequency.

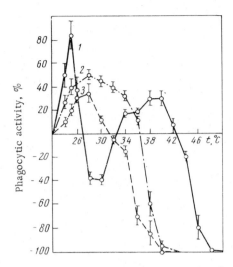

Fig. 61. Change in phagocytic activity of paramecia due to SHF field (1), infrared irradiation (2), and convective heating (3).

9.4. Effects at the Molecular

Level *in Vitro*

An effect of SHF fields on the structures of biological molecules was first detected in a series of investigations by Van Everdingen (1938, 1940, 1941, 1946b, 1946c) with solutions of glycogen, starch, and animal tissue extracts in carbon disulfide.

Exposure of an aqueous solution of glycogen to SHF fields of 1875 and 3000 MHz led to a reduction in the angle of rotation of the plane of polarization in this optically active solution. The magnitude of this effect and the nature of its development during irradiation depended on the frequency of the EmF, the concentration of the solution, and its viscosity. The frequency of 3000 MHz was much more effective and, hence, it was used in most of the experiments. The effect depended very considerably on the concentration and viscosity of the glycogen solution, as Fig. 62 illustrates.

The effect of a SHF field on the structure of colloidal solutions of starch was manifested in a pronounced increase in the rate

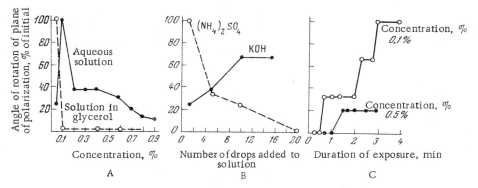

Fig. 62. Change in angle of rotation of plane of polarization in glycogen solution due to SHF field in relation to concentration of solution (A), number of drops of KOH or $(NH_4)_2SO_4$ added to solution (B), and duration of exposure (C).

of coagulation (estimated from the precipitation) in 0.1-0.2% solutions. An opposite (in comparison with glycogen solutions) effect of a SHF field was observed in experiments with optically inactive extracts of skin and fat tissues (in carbon disulfide): These extracts became optically active (dextrorotatory) as a result of irradiation.

The results of Van Everdingen's experiments stimulated research on the effect of EmFs on human gamma globulin (Bach, 1961; Bach et al., 1961a, 1961b). In these investigations methods previously employed for detection of the effect of x rays on gamma globulins were used. The effect was detected by the change in the electrophoretic pattern from a single peak to a double peak and by measurements of the antigenic reactivity of the gamma globulin (by titration against the serum of a rabbit immunized against human gamma globulin). A solution of human gamma globulin (in 2.2% physiological saline) was placed in a chamber with flat electrodes to which a voltage was applied either from a pulse generator with a frequency range of 10-200 MHz (with pulse lengths of 10-60 μsec and a pulse repetition rate of 500-2000 per sec) or from a continuous generator providing the same frequencies. The duration of exposure was 20-30 min. Cooling of the electrodes with water during irradiation kept the solution temperature in the range 30-40°C (recorded by a thermocouple).

In the first series of experiments the electrophoretic patterns of the solutions after exposure to EmFs of intensity 60 mW/cm^2 and frequencies at every 10 MHz over the whole frequency range were investigated. The electrophoretic pattern was altered from a single peak to a double peak at certain frequencies: 30, 60, 140, 180, and 200 MHz. Fuller investigations in the "active" range in the region of 30 MHz. (at 1-MHz intervals) showed that this effect was most pronounced at frequencies of 29, 31, and 34 MHz. The frequency dependence was most marked in the range 13-13.34 MHz, where the effect occurred at frequencies of 13.1, 13.12, 13.2, 13.3, and 13.32 MHz.

Even more interesting results were obtained when the effect was assessed by titration of irradiated and nonirradiated solutions of the gamma globulin of serum of a rabbit (immunized against human gamma globulins). This revealed "effective" frequencies at which the irradiated solutions showed much higher titers than nonirradiated solutions or solutions exposed to other frequencies. The effective frequencies were either the same as the frequencies revealed in the electrophoretic investigations or differed from them by a few tens of kHz. Figure 63 shows the results of these investigations with different irradiation conditions. It is obvious from these results that the effects are manifested at low intensities and that the main factor is the field strength, and not the amount of energy absorbed in the solution.

Investigations in vitro have revealed changes in enzyme activity due to EmFs of various frequency ranges and also to constant magnetic and constant electric fields.

In view of the above-discussed (Section 4.5) hypothesis on dipole—dipole interactions between enzyme and substrate molecules, Vogelhut (1960) carried out experiments with the enzyme lysozyme. Exposure to a SHF field of a particular resonant frequency (in the range 8200-12,400 MHz) altered the optical density of the solution and the rate of the enzyme—substrate reaction.

A reduction of the activity of the enzyme alpha-amylase due to UHF fields of specific frequencies has also been observed (Bach, 1965). The enzymatic activity was assessed from the optical density of the solution. It was found that the effect depended on the temperature and that the greatest reduction in activity occurred in the range 27.7-28.2°C. Figure 64 shows samples of the auto-

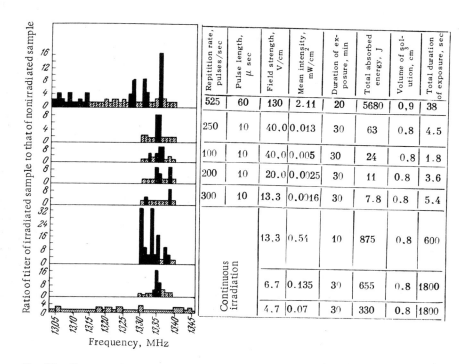

Repitition rate, pulses/sec	Pulse length, μ sec	Field strength, V/cm	Mean intensity, mW/cm²	Duration of exposure, min	Total absorbed energy, J	Volume of solution, cm³	Total duration of exposure, sec
525	60	130	2.11	20	5680	0,9	38
250	10	40.0	0.013	30	63	0.8	4.5
100	10	40.0	0.005	30	24	0.8	1.8
200	10	20.0	0.0025	30	11	0.8	3.6
300	10	13.3	0.0016	30	7.8	0.8	5.4
Continuous irradiation		13.3	0.54	10	875	0.8	600
		6.7	0.135	30	655	0.8	1800
		4.7	0.07	30	330	0.8	1800

Ratio of titer of irradiated sample to that of nonirradiated sample

Frequency, MHz

Fig. 63. Change in human blood gamma-globulin activity due to UHF fields of particular frequencies.

matic recording of the variation in optical density of the solution for different rates of sweeping of a frequency of about 11.8 MHz. A change in alpha-amylase activity due to a radio-frequency EmF has also been reported (Korteling et al., 1964).

A change in the activity of the enzymes catalase and peroxidase due to exposure to UHF and low-frequency EmFs was found in experiments with diluted (1:1000) rabbit blood (Chirkov, 1964, 1965). Exposure to an EmF (50-72.5 V/cm, 20 min) led to a reduction in the activity of both enzymes, but the frequency dependence was different for the investigated ranges. In the low-frequency range it was of a resonance nature — the reduction occurred only at frequencies which were multiples of 4 kHz and was particularly pronounced at 8 and 20 kHz (Fig. 65A), whereas in the UHF range there was a gradual increase in the effect with increase in frequency (Fig. 65B).

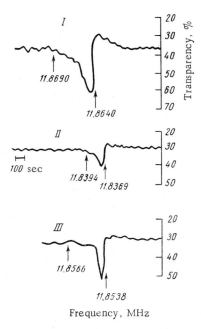

Fig. 64. Curves showing changes in optical density of alpha-amylase solution due to an EmF of 11.8 MHz. I) Sweeping of frequency at a rate of 6 kHz in 10 min; II) 6.2 kHz for 10 min; III) 5.7 kHz for 10 min.

Cook et al. (1964) investigated the effect of a magnetic field on enzymatic activity in experiments with trypsin solutions by comparing the absorption spectrum in the ultraviolet region before exposure to the field. It was found that exposure to 8000 Oe reduced the absorption at 253.7 mμ (Fig. 66). A reactivating effect of a magnetic field of 5000 Oe on trypsin partially inactivated by autolysis at pH 7–8 or by egg-white inhibitor has also been observed (Wiley et al., 1964). Such a field, however, had no effect on trypsin inactivated by soybean inhibitor, diisopropylphosphorofluoridate, or ultraviolet irradiation.

A stimulating effect of a magnetic field of 20,000 Oe was found in experiments with the enzyme carboxydismutase (Akoyunoglou, 1964). After a 48- to 192-h exposure to the field the enzyme activity was increased by 14–20% (but inactivation of the enzyme by ultraviolet irradiation was not reversed by the magnetic field). Similar effects were produced by an electrostatic field.

The birefringence method recently revealed appreciable orientation of DNA molecules in a solution exposed to a magnetic field of 6500 Oe (Mekshenkov, 1965a, 1965b). It was found that the DNA molecule (its rigid segment) oriented itself with its long axis perpendicular to the magnetic field lines.

From the viewpoint of the study of the biological action of a magnetic field at the molecular level the effect of such a field on the physicochemical properties of water is of great interest. It was discovered a considerable time ago that water which had been exposed to a magnetic field was less hard and produced less scale

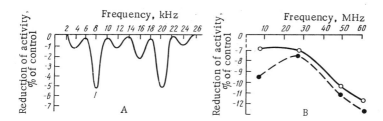

Fig. 65. Plots of reduction of catalase (continuous curves) and peroxidase (broken curves) activity in rabbit blood in vitro due to EmF against frequency. A) low-frequency range; B) UHF range.

than ordinary water. This effect has found practical application in steam boilers. The method of magnetic treatment is very simple: The water flows through a glass tube a few millimeters in diameter at a fairly high velocity (0.3–0.6 m/sec) between the poles of a magnet or a series of magnets.

Investigations conducted in recent years have shown that other properties of water are altered by a magnetic field; these effects depend on the exposure conditions (Minenko et al., 1962). The effect of a magnetic field on the amount of deposit was determined by evaporation of the water (to 0.2 of its initial volume) and weighing the deposit (scale) and suspended particles. The effect (A) was assessed from the relationship

$$A = \frac{a_0 - a}{a} \cdot 100,\qquad(43)$$

where a_0 and a are the weights of scale from the untreated and treated water, respectively.

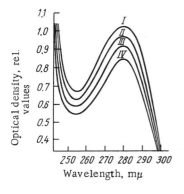

Fig. 66. Reduction of ultraviolet absorption in trypsin solution due to a magnetic field of 8000 Oe. I) Control; II) 2 h in field; III) 5 h in field; IV) 7 h in field.

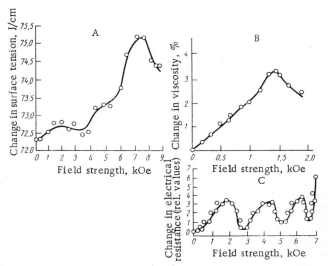

Fig. 67. Change in physicochemical properties of water due to magnetic field in relation to field strength. A) Surface tension; B) viscosity; C) electrical resistance.

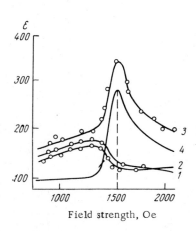

Fig. 68. Change in dielectric constant of water due to magnetic field. 1 and 2) Still distilled water and tap water, respectively; 3) flowing distilled water; 4) difference in curves 3 and 1.

The investigations showed that the observed effects (reduction of hardness and amount of scale) were greatest at two "optimum" field strengths — about 1500 and 4500 Oe.

The reduction of the amount of scale depended also on the speed of flow of the water in the magnetic field and was greatest in the range 0.4–0.5 m/sec; at lower and higher speeds the effect was weaker.

A crystal chemical analysis showed that the scale and suspended particles in water treated with a magnetic field consisted of crystals of rhombohedral and other shapes, whereas needle-shaped crystals

usually predominated. A very important component part of the
scale and suspended particles was calcium carbonate, which is usually released in the form of the stable crystalline modification
calcite; the treated water contained a certain amount of the unstable modification aragonite.

Changes in the physicochemical properties of water due to a
magnetic field have also been investigated in experiments with distilled water. These experiments showed an increase in surface tension, viscosity, and electrical conductivity of the water, and again
the effects were most pronounced at one, two, or even several "optimum" field strengths (Fig. 67).

The effect of a magnetic field on the dielectric constant of
water (Umanskii, 1965) is shown in Fig. 68. The greatest effect is
again found at 1500 Oe.

Bruns et al. (1966) recently investigated the effect of a magnetic field on the absorption of light (conditions of treatment: tube
6 mm in diameter, speed of flow 0.6 m/sec, nine electromagnets).
The change in the absorption of light by the water (in relative units)
was measured 10 min after its exposure to the magnetic field.
The field strength was varied in the range 0–1500 Oe and was evaluated from the current. The graphs in Fig. 69A show that at all
the employed field strengths the absorption maximum occurred at
the same wavelength. This means that the magnetic field does not
lead to dissociation or association of molecules. The plot of absorption against field strength (Fig. 69B) has two optima.

Of great interest from the viewpoint of the biological action
of a magnetic field are the experimental data which indicate that
water treated by a magnetic field can affect the behavior and vital
activity of living organisms.

We have already mentioned that paramecia in a capillary
containing water which has been placed for some time in a magnetic field collect at the end facing the south magnetic pole (Kogan
and Tikhonova, 1965).

Experiments with mice (Glebov et al., 1965) showed that when
the animals were given an intravenous injection (1 ml per 20 g of
weight) of water which had been exposed to a magnetic field of
1000–1500 Oe there was increased urination in 63% of cases in com-

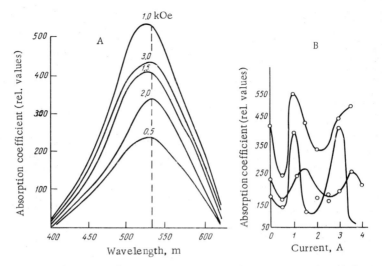

Fig. 69. Absorption of light in distilled water flowing in a magnetic field. A) Absorption spectra at different field strengths (indicated on curves); B) absorption in relation to current, i.e., field strength (three different experiments).

parison with animals which had received an injection of ordinary water. In mice which had human kidney stones grafted into their bladder and then received 1 ml of magnetically treated water daily (in milk) the number of crystals formed around the stones after 30 days was only two-fifths of the number in the control (P = 0.015). In mice which received magnetically treated water in the morning the adrenals were enlarge by 20% and the spleen was reduced by the same amount (P < 0.05). Experiments with rats (Dardymov et al., 1966) showed that water treated with a mgnetic field affected the resistance of erythrocytes to alkaline hemolysis; in adult animals which received such water for a month hemolysis was 20-40% more rapid than in the control.

An effect of magnetically treated (1000–1500 Oe) water on plants has also been discovered (Dardymov et al., 1965). Watering of sunflower, corn, and soybean seedlings with such water (100 ml per plant per day) led to accelerated growth (Table 15). In a second series of experiments the effects of magnetically treated distilled and tap water on the development of soybean were compared. In each case there was a significant (P < 0.05 and < 0.001) acceler-

TABLE 15.   Effect of Distilled Water Exposed to a Magnetic
Field of 1000 Oe on the Growth of Some Plants by the 12th
Day After Sowing

Plant	Experimental group	Height of plants, mm		Thickness of stem, mm	
Sunflower	Control	38.4 ± 0.58	$P < 0.001$	8.0 ± 0.0	$P > 0.05$
	Experiment	46.7 ± 0.29		9.28 ± 0.93	
Corn	Control	22.0 ± 16.8	$P > 0.05$	2.3 ± 0.21	$P < 0.01$
	Experiment	41.2 ± 1.12		6.0 ± 0.58	
Soybean	Control	53.35 ± 11.27	$P < 0.01$	4.0 ± 0.47	$P > 0.05$
	Experiment	74.5 ± 11.27		4.15 ± 0.11	

ation of the growth of the plants in comparison with the control and
this effect was more pronounced in the case of tap water.

Information on the effect of high-frequency EmFs on the properties of water has been obtained (Plaksin et al., 1966): Exposure
to fields of frequencies from 100 kHz to 8 MHz for 30 min increased
the optical density of water in the range 380-691 mμ.

Guinea pigs and mice which received only water which had
been exposed to an EmF of frequency 10 kHz (Valfre et al., 1964)
showed downward fluctuations in weight in comparison with animals
which received ordinary water. A similar effect was noted in the
second generation of the animals, even in cases where these progeny received ordinary water. The authors pointed out that the
effect of the treated water was particularly pronounced in periods
of solar activity.

A comparison of the effects of EmFs at the molecular level
in vitro and in the entire organism show that they differ significantly. The nature and magnitude of changes at the molecular level in the organism are almost independent of the frequency, whereas in experiments in vitro such effects are frequency-dependent and
are often of a resonance nature. The effect of EmFs on the biochemical activity of macromolecules in vitro may even by the opposite of that observed in the entire organism. There is also a difference in the relationship between the effects and the EmF inten-

sity: The two-optima and two-phase relationship in the entire organism is not found in experiments in vitro, except in the case of the action of a magnetic field on water, where such a relationship is found.

This indicates how risky it is to try to determine the mode of action of EmFs on biochemical processes in the organism from investigations of such action in experiments in vitro.

Chapter 10

MECHANISMS OF THE EXPERIMENTALLY OBSERVED BIOLOGICAL EFFECTS OF ELECTROMAGNETIC FIELDS

The phrase "mechanism of the biological action of a particular factor" ususally means the prime cause of the reaction of the biological object to the factor. The elucidation of the mechanism of the biological effect is usually taken to mean the establishment of a cause-and-effect relationship between the factor and the reaction, to determine the systems or structures involved in this reaction, and to discover the processes responsible for the reaction.

It is often asserted that a complete explanation of the mechanism of the action of any factor on the organism necessitates the determination of the associated physicochemical processes at the molecular level, which are ultimately responsible for the reaction of the organism. In several cases such p h y s i c o c h e m i c a l m e - c h a n i s m s can actually be found by investigating the effect of the particular factor on isolated organs and cells and on molecular processes in vitro.

However, many of the changes in the organism due to its reaction to an external factor can be detected only at the macroscopic level in fairly complex systems, while the molecular processes underlying these changes remain quite obscure. On the other hand, in cases where effects at the molecular level are found, they are not necessarily the primary cause of the changes occurring at higher levels of organization, since such effects themselves are frequently the result of changes in complex systems. We have already mentioned that attempts to elucidate the mechanism of action of an external factor on the organism from the results of investigations of the effect of this factor on isolated organs and tissues and on molecular processes in vitro are often quite unjustified, since the ability to react in a particular way to a certain factor may be a feature only of a system of a particular level of complexity and the components of this system may either react differently to this factor or not react at all to it.

This is illustrative of a specific characteristic of living organisms — their hierarchic organization. As we pointed out in the introduction, each stage of complexity in the hierarchy of the systems of the organism has its own specific properties, which determine its interaction with other systems of the hierarchy and its reaction to external factors. It is difficult, and often impossible, to predict these properties from an investigation of systems of a lower degree of organization or the elements comprising the considered system. Hence, the investigation of the mechanisms of the reactions of an organism to external factors must begin at the level of the entire organism and then proceed to the least complex level at which the changes ultimately responsible for the over all reaction can still be detected.

The first stage in such an investigation of mechanisms is to determine the qualitative and quantitative aspects of the cause-and-effect relationship between the external factor and the reactions of the organism as a whole and its systems of different complexity to it. Such relationships can usually be described only phenomenologically. Hence, in scientific literature we frequently encounter such terms as physiological mechanisms of the action and biophysical mechanisms of the action. These concepts express the relationship between the "input," i.e., the factor, and the "output," i.e., the reactions, in systems of organs, individual organs, tissues, and cells. For instance, physiological mechanisms of action include vagotonic or sympothotonic reactions, excitation or inhibition in the central nervous system, alteration of the excitability of nerve and muscle tissues, and so on. Biophysical mechanisms include changes in the electric parameters of tissues or in the nature of the bioelectric processes occurring in them, changes in the permeability of cell membranes, and so on.

The second stage in the investigation of mechanisms is to simulate the systems and processes involved in the reaction of the organism to the external factor. A model of a biological system or process is usually a mental picture of the structure and connections of the investigated phenomenon or process. As we have already pointed out, simulation in biology is rarely effective if it is based only on the phenomena and principles characteristic of non-living matter. Comparisons with structural and functional schemes based on "organized" machines are more valid. Of course, any form of simulation entails the inevitable simplification of the investigated phenomenon and does not give a definite answer to the question of the cause-and-effect relationships underlying it. However, by starting off with some hypothetical scheme consistent with the main phenomenological features of the biological effect, we can proceed by the method of successive approximations, corrected by indirect experimental data, to obtain a more accurate picture of the investigated effect and, in particular, of the mechanism of the biological action of the external factor.

Finally, in cases where information relating to the physico-chemical mechanisms involved in the investigated reaction of the biological object to external factors is available the investigation of the mechanism can be pursued to the implicated processes at the molecular level.

Such a general methodological approach can also be adopted in the investigation of the mechanisms of the biological effects of EmFs. As we have seen, the nature of the reaction of the animal organism to EmFs depends (when the exposure conditions are the same) on which systems of the organism are acted on directly and on whether these systems are in the entire organism or isolated. In several cases the effects observed at the molecular level in the entire organism cannot be found in experiments in vitro.

10.1 General Considerations in Simulation of the Mechanisms of the Biological Effects of EmFs

We will consider briefly the basic features which have been discovered in experimental investigations of the biological effects

of EmFs and which can be of assistance in simulation of the mechanisms of these effects.

First of all, we note that in most cases EmFs cause d i s -
t u r b a n c e s in the regulation of physiological processes. Such
disturbances are particularly pronounced during embryonic development and during growth, i.e., in the period when the defense
mechanisms do not exist or are not fully developed. An examination of the nature of these disturbances indicates that they are due
to the effect of the EmF on the electromagnetic processes involved
in the regulation of physiological functions.

In most experimental investigations the choice of the EmF
parameters has been determined by the generator available to the
experimenter rather than on the basis of biological considerations.
It has been only in certain cases that the EmF parameters have
been chosen on the basis of assumed parameters of the biological
systems. Hence, the parameters of the acting EmFs have usually
been inappropriate to the biological system, and such action can be
regarded as i n t e r f e r e n c e . We use this term in accordance
with Salzberg's definition (1966): "Interference is an undesirable
signal in a particular system, which in another system is useful
or coherent with the useful signal in another system."

It is obvious that each living organism must be reliably protected from external natural and artificial electromagnetic interference (or n o i s e — signals which are not coherent with any useful signal of the system). The multistage system of protection of the
organism probably accounts for the experimentally discovered two-
optimum (or many-optimum) and two-phase relationship between
the biological effects of EmFs and their intensity. This defense
system evidently accounts also for the fact that the disturbances
produced by EmFs of different frequencies are often of the same
type.

There are probably two types of defense system — passive
and active. The first, a rapidly responding peripheral system, can
act by means of appropriate frequency and amplitude filters and
correlators; each successive stage in such a defense system is
brought into action by a certain range of the parameters of electromagnetic interference (the main parameter is the intensity).
This passive protection could account for the two-optimum (or

many-optimum) relationship between the disturbances and the intensity of the acting EmFs. When the passive defense system becomes inadequate the active system comes into action: The rapidly acting peripheral system signals the arrival of interference to the slowly reacting central system (which regulates physiological functions) and the latter actively protects itself by reducing its sensitivity to the reception of interference. Another, higher stage of active defense is the central system, which regulates physiological processes and modifies them in such a way as to counter the undesirable external action on the organism.

Such a process could account for the appearance of opposite physiological changes due to EmFs of low and high intensities, i.e., for the two-phase relationship between the biological effects of EmF and their intensity.

The correctness of these model schemes is confirmed by experimental information regarding the reactions of the animal organism to the action of EmFs on the central or peripheral nervous system or in the case of whole-body exposure, which affects both these systems (see the summary at the end of Chapter 7). These schemes are also consistent with the results of experiments on animals like unicellular organisms, which have no nervous system but have a "central" intracellular control system and a peripheral system in the cell membrane.

The reason for the cumulative effect of EmFs must be sought in the ability of biological systems to store "useful" information, as well as interference or noise: In the discussed evidence of the cumulative effect of EmFs we encountered directly detectable changes in biological structures, as well as effects due to an accumulation of functional changes.

We turn now to a discussion of the mechanisms of the biological effects of EmFs in cases where their analysis can be taken to processes at the molecular level and in cases where the effects are obviously due to the action of the EmF on macromolecules, extracellular and intracellular media, and so on. Some ideas on this, applicable to different types of biological effects of EmFs, have already been put forward (Presman, 1963, 1965a, 1966a, 1966b, 1967a).

10.2. Mechanisms of Action of
EmFs on Neurohumoral Regulation

As already mentioned, the experimental data indicate that the reason for the disturbance of the regulation of physiological functions in animals by EmFs is the direct action of EmFs on various divisions of the nervous system. A comparison of the observed effects suggests that the enhancement of the excitability of the central nervous system is of a reflex nature due to the direct action of EmFs on the peripheral divisions, while the inhibiting reactions are usually associated with direct action on the structures of the brain and spinal cord. There are grounds for postulating that the cerebral cortex and interbrain structures, particularly the hypothalamus, are most sensitive to EmFs. It is obvious that the effect of EmFs on the nervous system is due either to stimulation of the nerve cells or to a change in the parameters of their functional state — excitability, amplitude of biopotentials, speed of conduction of excitation, and so on.

What physicochemical processes underlie such effects of EmFs on nerve cells? There is still not enough theoretical and experimetal information to answer this question. We still actually know very little about the physicochemical nature of nervous excitation itself, although there is a vast amount of experimental data on the physicochemical processes asociated with it. Yet some ideas regarding the possible physicochemical mechanisms of action of EmFs on nerve cells can be derived from a consideration of the main features of this action and modern ideas on the processes occurring when a nerve is electrically stimulated. In view of the experimentally established broad-band (as regards frequency) sensitivity of nerve cells to EmFs the following mechanisms are probable (Presman, 1962a, 1963e):

1. EmFs can be detected in the membrane of the nerve cell (Cole, 1950) and, hence, at different frequencies there will be a direct component of rectified current which will lead to stimulation of the cells or a change in their excitability.

2. EmFs can affect the mobility of ions, which are implicated in the process of nerve excitation. Vibrations of ions at the fre-

quency of the acting field will have some effect on their ability to penetrate the membrane of the nerve cell and, hence, on its excitability.

3. According to current ideas (Samoilov, 1957), the thermal and, in particular, translational motion of water molecules is more difficult in the presence of sodium ions than in pure water (positive hydration), whereas in the presence of potassium ions water molecules are more mobile than in pure water (negative hydration). The translational motion of the ions themselves is due to transfer of the water molecules closest to them.* In view of this, we can postulate that the action of EmFs (of sufficiently high frequencies) on the water molecules surrounding sodium and potassium ions will be different and, hence, the change in the mobility of these ions will be different. This in turn can lead to a change in the potassium— sodium gradient between the cell and the extracellular medium and, hence, to excitation or a change in the excitability of the nerve cell.

4. The effect of EmFs on the permeability of the nerve cell membrane could also be attributed to induced vibrations of the water molecules hydrating the protein molecules in the surface layer of the membrane. This would lead either to excitation of the nerve cell or to a change in its excitability.

5. EmFs may also affect the so-called spontaneous activity of receptors; the nature of this effect is still obscure, but it is obviously associated with thermal disturbances of ionic processes in the membrane. Vibrations of the ions at the frequency of the acting field may introduce some "order" into these chaotic processes and, hence, alter the nature and the intensity of spontaneous activity.

6. The cumulative effects of EmFs on the neurohumoral system can probably be attributed to the ability of various elements of this system (as of other biological systems) to store information or noise.

The mechanisms of action of EmFs on the regulation of physiological processes in the entire organism may be approached from other standpoints on the assumption that EmFs affect the internal electromagnetic regulation, which acts in different frequency

---

* This may be a fruitful concept for elucidation of the mechanism of the different permeability of the nerve fiber membrane for potassium and sodium.

ranges. In the light of this hypothesis EmFs can upset the function of a particular link in this regulation by increasing its activity in cases where the frequency of the acting EmFs is close to the natural frequency of the particular structure or by reducing it when the structure is forced to vibrate at uncharacteristic frequencies. This approach to the mechanisms of the biological action of EmFs will be discussed more fully in the third part of the book, where electromagnetic regulation systems in living organisms are discussed.

10.3. Mechanisms of Action of EmFs on the Reproduction and Development of Organisms

The experimental data discussed in Chap. 8 show that EmFs have a particularly pronounced effect on reproduction and embryonic development and on the growth of young organisms. This is quite understandable, since any disturbances in the control of biological processes due to EmFs (irrespective of the mechanisms responsible for these disturbances) will be more likely to occur at the stages of formation of the organism, when the defense mechanisms are still undeveloped or are incompletely developed.

As we have seen, the effect of EmFs on reproductive processes can in several cases be attributed to the disturbance of neurohumoral regulation in the organism and consequent alteration of the functions of the endocrine system, cell metabolism, and enzymatic processes. However, some effects appear to be due to the direct action of EmFs on the functions of germ cells or to irreversible changes in these cells due to EmFs. Here we encounter the effect of EmFs on physicochemical intracellular regulation processes and the destruction of intracellular structures. Experiments in vitro confirm that EmFs actually do act on intimate processes in cells.

The effect of EmFs on intracellular processes evidently accounts for the genetic effects, which can be due either to disturbances of biochemical processes in the cytoplasm or to the direct action of the EmF on DNA. The possible mechanism of such an effect is worth discussing.

It was recently shown that the interaction of homologous portions of chromosomes might be due to the electric field produced
by fluctuations in electric charge density (Zyryanov, 1961). Yet,
as already mentioned, it has been shown experimentally and theoretically that interactions can take place between macromolecules
due to radio-frequency EmFs as a result of dipole—dipole transitions due to fluctuations of the charge distribution in the molecules
(Vogelhut, 1960). If dipole—dipole transitions of this kind occur
also in DNA molecules, a resonance effect of EmFs of particular
frequencies on these molecules, as well as disturbances of the interactions between such molecules (or between them and RNA molecules) due to EmFs, could be expected.

The mechanisms of the mutagenic effect of a magnetic field
might be approached from the standpoint of the theoretically possible (Dorfman, 1962) orientation of DNA molecules in the field.
Since a magnetic field penetrates the animal's entire body, then
this field can directly affect the macromolecular processes in the
entire organism. This could also be the cause of the impairment
of reproductive processes and inhibition of the growth of animals
in a magnetic field.

Direct action on molecular processes could also be responsible for the effects of SHF fields on the development of the chick
embryo and on the cardiac activity of this embryo. Finally, the effect of the magnetic field on plant development is also probably due
to the orienting action on macromolecules.

The greatest amount of information for elucidation of mechanisms of direct action of EmFs on macromolecules can, of course,
be obtained from an analysis of the in vitro effects of EmFs at the
cellular and molecular levels.

10.4. Mechanisms of Action of EmFs on Unicellular Organisms and Cell Cultures

As already pointed out, the effects of EmFs on unicellular
organisms are of two types. On one hand, there are the orientational
and motor reactions of unicellular organisms, which can be ascribed
to the action of EmFs on peripheral excitable structures; on the
other hand, there are the reactions which appear to be due to the

effect of EmFs on intracellular control processes. At present there is experimental evidence only for discussion of the "peripheral" action of EmFs.

The possibility of detection of an EmF in the cell wall of paramecium has been confirmed (Presman, 1963e; Zubkova, 1967a) in comparative experiments in which these infusoria were stimulated by pulses of alternating current, rectified (single half-cycle) alternating current, and series of dc pulses of the same frequency and off-duty factor. The amplitude of the threshold voltage for the motor reaction was the same for all these kinds of treatment.

An attempt was made (Zubkova, 1967a, 1967b) to explain the mechanism of action of EmFs on the excitable structure of paramecia by the enzyme theory of nerve excitation. According to this theory, excitation is initiated by the reaction of acetylcholine with a receptor protein; this enhances the permeability of the membrane and "triggers" processes which lead to rapid migration of sodium and potassium ions. The acetylcholine — receptor complex is in dynamic equilibrium with the free components of this complex and free acetylcholine is rapidly inactivated by the enzyme cholinesterase. The application of this theory to paramecia is quite justified since it has been known for a long time that the pellicle (outer membrane) contains an acetylcholine — cholinesterase system and it has been shown that this system is implicated in the excitation process.

Zubkova measured the threshold voltage of the alternating current at which the motor reaction (electric shock reaction) occurred in paramecia in the following conditions: a) In a medium containing proserine, which blocks the action of cholinesterase and reduces the excitability of paramecia, or cysteine, which acts on protein sulfhydryl groups and also reduces the excitability; b) in media containing these substance and exposed to a SHF field, which increases excitability, c) in an ordinary medium exposed to a SHF field. The graphs in Fig. 70 show that a SHF field counteracted the effect of both proserine and cysteine. In the first case this was possibly due to restoration of the upset dynamic equilibrium between free acetylcholine and cholinesterase, and in the second case it could have been due to inhibition of oxidation of sulfhydryl groups. Thus, SHF fields apparently act on both links of the excitation-initiating reaction.

In the discussion of the action of EmFs on cell cultures we encountered a frequency dependence. This can be attributed to the direct action of EmFs on subcellular structures or on macromolecules, which react to resonant frequencies.

A hypothesis to explain the inhibition of cell growth by a magnetic field was recently put forward (Liboff, 1965). It was suggested that the active transport of metabolites (in ionic form) may be effected by the electrostatic field of the uniformly charged cell nucleus or cytoplasm. In a magnetic field the ion diffusion coefficient is reduced. If the cell is assumed to be cylindrical and the electric field is produced by cytoplasmic charges, then a magnetic field parallel to the cell axis will reduce the velocity of migration of ions. A detectable effect can be expected at a field strength of the order of 10^5 Oe.

10.5. Mechanisms of Action of EmFs at the Molecular Level

We have already expressed some theoretical views on the possible mechanisms of action of EmFs at the molecular level. Experimental investigations in vitro have shown mainly resonance effects and these in a very wide frequency range — from low to superhigh. Yet in recent years more and more new mechanisms of resonance absorption of EmFs in biological media have been discovered.

Shnol' (1965, 1967) showed in a series of investigations that vibrational processes with infrared to audio frequencies take place in solutions of actomyosin, actin, and myosin. These are presumably conformational vibrations of protein molecules, consisting in the formation of folds, twisting, or compression of the polypeptide chains. The concomitant changes in the hydrophilic — hydrophobic properties of the surface of protein moleculas cause a corresponding reorganization of the structure of water. "Hydrophilic — hydrophobic waves" are propagated in the water and synchronize the vibrations in the macrovolume of the protein.

Conformational vibrations of protein molecules lead to the displacement of the electric charges on their surface and, hence, EmFs of particular frequencies can interact with these vibrations. This mechanism might underlie the above-described resonance

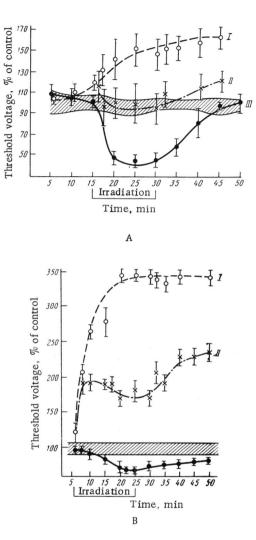

Fig. 70. Variation with time of threshold voltages of
ac pulses causing the electric shock reaction in paramecia in various conditions. A: I) Addition of proserine $(1.5 \cdot 10^{-3}\%)$ to medium; II) addition of proserine and exposure to SHF field; III) exposure to SHF
field. B: I) Addition of cysteine (1%) to medium;
II) addition of cysteine and exposure to SHF field;
III) exposure to SHF field. The hatched areas are the
threshold voltages in the control.

effects due to the action of audio-frequency EmFs on catalase solutions.

In another investigation Chernavskii et al. (1967) examined the much higher frequency vibrations of enzyme molecules due to their elastic deformation. According to the authors' theoretical estimate, the frequency of vibrations for molecules of the enzyme lysozyme (the structure of which has been completely worked out) is $1 \cdot 10^{10}$ to $5 \cdot 10^{10}$ Hz.

It is of interest that in the above-described experiments on the effect of SHF fields on a solution of lysozyme and its substrate, Vogelhut (1960) found a resonance effect (change in rate of enzyme—substrate reaction) at a frequency of 10^{10} Hz. It is to be hoped that an examination of the elastic deformational vibrations of protein molecules might lead to an elucidation of the mechanisms and other resonance effects of ultrahigh-frequency EmFs on gamma globulins.

Another type of resonance absorption in the radio-frequency region of the spectrum, which the authors called "piezoelectric resonance," has been observed in some biopolymers (Tul'skii et al., 1965). In the 15 to 50MHz range he found a large number of resonance lines 2 to 200 kHz broad, the position and intensity of which depended on the temperature. Unlike the signals due to nuclear quadrupole resonance the absorption lines are not affected by a constant magnetic field and are due to the electric component of the EmF. When the temperature of the specimen changes the frequency of the lines varies in accordance with the law

$$ -\frac{1}{\nu} = \frac{d\nu}{dT} = \Delta, \tag{44} $$

where ν is the frequency of the absorption line; T is the temperature of the specimen; and Δ is a constant characteristic of the substance. This constant has been measured for various substances, including biological substances.

The authors put forward the hypothesis that piezoelectric resonance is due to the interaction of an elastic wave produced by the EmF, with defects and inhomogeneities on the surface and in the interior of the substance. Elastic waves are produced in substances which have at least small regions with piezoelectric properties. In view of this the authors believe that if the effects observed in biopolymers are actually of a piezoelectric nature, then biological structures must contain regions consisting, possibly, of many macromolecules with piezoelectric properties. Thus, here we encounter the mechanism of action of EmF on macromolecular ensembles.

Water must play a significant role in the effect of EmFs at the molecular level (in vitro and in vivo). It has been suggested (Klotz, 1962) that protein molecules in an aqueous solution make the water molecules form a stable cystalline structure — crystal hydrates (hydrotactoids) are formed around the nonpolar groups of the protein molecules. Denaturation of the protein molecules (by heat treatment, urea, etc.) is presumably due to the destruction (melting) of the hydrotactoids. There is also experimental evidence that water is implicated in the formation of the specific structure typical of the macromolecule, i.e., plays a significant role in stabilization of the structure of protein molecules. The effect of EmFs on protein molecules may be largely due to their action on water molecules and microcrystals and, hence, on the ordering of the hydrate shells of protein molecules, on crystal hydrates in these molecules, and so on. It is known that the characteristic relaxation frequency for free water molecules is in the SHF range, and for ice is in the audio-frequency range (it is assumed that liquid water has the same hexagonal structure as ice).

Thus, there are theoretical and experimental grounds for approaching the mechanisms of the effects of EmFs at the molecular level by considering processes in a very wide range of structures — from macroscopic processes involving ensembles of molecules (as in conformational vibrations of protein molecules and in piezoelectric resonance) to processes involving the orientation of nuclear spins. However, as we have already stressed repeatedly, between the molecule and the cell there is a hierarchy of structures of increasing complexity, and processes at the molecular level in the cell and, even more so, in the organism, are different from those occurring in experiments in vitro. All this must be considered in any attempts to explain the mechanisms of the action of EmFs on molecular processes in living organisms.

SUMMARY

The experimental data discussed in the second part indicated quite
clearly the extreme diversity of the biological effects of EmFs on
diverse representatives of the animal and plant kingdom and on all
biological systems — from macromolecules to the entire organism —
have been observed.

A very important feature of the biological effects is that they
are often produced by fields of extremely low intensities. This naturally suggests that living organisms have systems which are particularly sensitive to EmFs and as yet have no analogs in engineering.
Such systems could have been formed in the course of evolution of
life by the constant interaction with environmental EmFs. A comparison of the experimentally determined sensitivity of living organisms to EmFs with intensities of the corresponding natural
EmFs shows that this is very probable. It should be borne in mind
that these are not necessarily the highest sensitivities inherent to
living organisms. Data on these sensitivities have been obtained in
experimental investigations with randomly selected EmF parameters differing in general from those to which organisms and their
various systems have become "tuned" in the course of evolution.

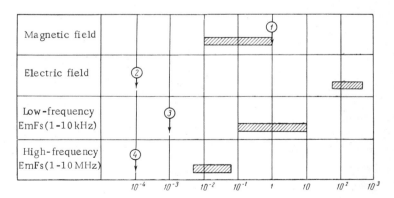

Fig. 71. Intensities of natural EmFs (hatched rectangles) and minimum
threshold intensities of reaction of biological systems to EmFs of corresponding frequencies. 1) Increase in motor activity of birds; 2) conditioned reflexes in fish with electric organs; 3) conditioned reflexes in
fish without electric organs; 4) vascular conditioned reflex in man.
The figures on the bottom line indicate the strength of the magnetic
field (Oe), electric field, or EmF (W/m).

With this stipulation we consider the diagram in Fig. 71. This figure shows the intensity ranges of natural EmFs in various frequency ranges and the minimum threshold intensities at which reactions of biological systems to EmFs of particular frequencies have been experimentally discovered. As the diagram shows, biological systems are sensitive to EmFs which have an intensity close to that of natural EmFs or even much weaker.

Thus, the discussed experimental data on the biological action of EmFs provide indirect confirmation of the validity of the hypothesis, put forward in the introduction, of the important role of EmFs in the evolution and vital activity of organisms. In addition, abundant experimental evidence directly confirming such a role of environmental EmFs in life has been amassed in recent years. There is also indirect and direct evidence which gives sufficient grounds for postulating that EmFs play an important role in information interconnections within living organisms themselves and are implicated in the transfer of "bioinformation" between organisms. This will be dealt with in the next part of the book.

Part Three

ROLE OF ELECTROMAGNETIC FIELDS IN THE REGULATION OF THE VITAL ACTIVITY OF ORGANISMS

In this part of the book we discuss the experimental investigations and theoretical arguments which directly or indirectly confirm the existence in living nature of three kinds of informational interconnections effected by EmFs — the transmission of information from the external environment to organisms, informational interconnections within organisms, and exchange of information between organisms. These connections could have been formed only by the development in organisms, in the course of evolution, of electromagnetic systems capable of receiving, transmitting, and converting the information carried by electromagnetic fields. These systems, on one hand, must be "tuned" to the detection of regularly varying environmental EmFs, to the reception and transmission of specifically coded EmF signals within the organism and between organisms, and, on the other, must be reliably shielded from natural electromagnetic interference, such as irregular spontaneous variations in environmental EmFs.

In the examination of the experimental data on the biological action of artificially produced EmFs in the second part of the book we encountered evidence of the high sensitivity of biological system to EmFs with particular parameters (frequency, strength, and nature of modulation) and the existence of defensive systems for protection against the adverse effects of EmFs. However, such features are revealed much more clearly by observations of the effect of natural environmental EmFs on living organisms of the most diverse species. Such experimental investigations will be discussed in Chapter 11. Experimental data and theoretical arguments in support of the existence of EmF interconnections and regulation within organisms are discussed in Chapter 12, and questions relating to the problem of EmF interconnections between organisms are dealt with in Chapter 13.

In the last chapter we discuss the practical application of the results of investigations of the biological activity of EmFs.

Chapter 11

ENVIRONMENTAL ELECTROMAGNETIC FIELDS AND THE VITAL ACTIVITY OF ORGANISMS

Hippocrates discerned the same phases in the course of illnesses and atmospheric phenomena and regarded this as a manifestation of the universality of the laws of nature and of all life. However, as Piccardi rightly points out (1965): "Since the time of Hippocrates no one has considered the effect of cosmic factors on biological processes. Only obvious meteorological variables have been considered."

It was not until the first decades of this century that V. I. Vernadskii undertook the scientific treatment of the problem of the universal relationship between living nature and cosmic processes. In "Essays on the Biosphere" (1926) he wrote: "The creatures of the earth are the creation of a complex cosmic process, an essential and natural part of a harmonious cosmic mechanism in which, as we know, there is no fortuity."

11.1   Relationship between Biological Phenomena and Solar Activity and the Role of EmFs

In 1915 Chizhevskii began a systematic investigation of the relationship between several biological phenomena and cosmic variables (Chizhevskii, 1963, 1964). He found a parallelism between changes in solar activity and colloidal-electric changes in the

blood, lymph, and cell protoplasm of animals, the growth of cultures of some bacteria, and so on.

It was later discovered (Vel'khover, 1936) that diphtheria
bacilli in years of maximum solar activity became less toxic and
more like the closely related harmless saprophytic bacteria.
The reaction (metachromasia) indicating this effect is observed in
bacteria four to six days before the appearance of flares and spots
on the sun.

Beginning in 1935 Japanese scientists (Takata et al., cited by
Chizhevskii, 1964) observed a relationship between the rate of
clotting of human blood and solar activity. At the times of passage
of spots across the central meridian of the sun the index of blood
clotting rate was more than doubled. This effect was correlated
with the rotation period of the sun (27 days) and the 11-year cycle
of solar activity.

In 1958 Dr. Shul'ts in Sochi conducted observations on changes
in the blood in relation to solar activity (Shul'ts, 1964). From a
large number of investigations (14,100 cases) and a review of the
relevant published information he found that the total leucocyte
count was reduced in periods of solar activity, although the number
of lymphocytes increased, as the curve in Fig. 72 shows. Shul'ts
noted a characteristic geographical feature of this effect: It was
most pronounced in polar regions and was practically absent in
equatorial regions.

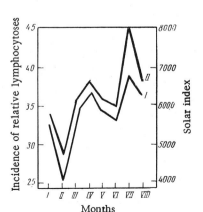

Fig. 72. Solar activity (I) and frequency of lymphocytoses (II) during
1957 in Sochi.

It has been known for a long time that the size of populations of a great variety of organisms increases in cycles with a period close to 11 years (Villee, 1952; Shcherbinovskii, 1964). This cyclicity has been observed in the growth of marine algae and coral colonies, and in the reproduction of fish, insects, and several mammals. Some scientists believe that the cause of such fluctuations is the 11-year cycle of solar activity. They point to the coincidence of plagues of the locust (Schistocerca) and periods of enhanced solar activity (Shcherbinovskii, 1964), to the correspondence between the thickness of the annual rings of Sequoia and solar activity (Chizhevskii, 1963), and so on. Research workers who do not subscribe to this view base their objections on the fact that by no means all cycles of increase in populations have the same periodcity as the sun and that a correspondence over a period of several years is followed by a divergence in which these maxima coincide with periods of minimum activity (Villee, 1952).

A correlation between enhanced solar activity and the incidence of various diseases was observed a long time ago. Such a relationship was first revealed in the thirties by Chizhevskii from an analysis of statistical data for many years (Chizhevskii, 1964). He showed that outbreaks of plague, cholera, influenza, diphtheria, and other infectious diseases coincided with periods of enhanced solar activity. Similar statistical investigations have been carried out on several occasions since and have given similar results. Figure 73 shows graphs indicating the relationship between solar activity and the number of cases of cerebrospinal meningitis and relapsing fever (Berg, 1960).

With what geophysical changes are all these biological effects in periods of solar activity connected? It is known that enhancement of solar activity leads to an increase in the intensity of visible and ultraviolet solar radiation, cosmic rays, ionization in the atmosphere, the strength of the earth's magnetic field, the intensity of atmospheric discharges, and so on. Experimental investigations carried out over the last 30 years, however, suggest that the main role in the considered effects is due to changes in envirommental EmFs, particularly magnetic and electric fields and atmospherics. We cite some experimental data on each of these kinds of biological effects.

An effect of the earth's magnetic field on the multiplication culture (R. Becker, 1963b). Four hours after inoculation the culture of bacteria was discovered in experiments with a Staphylococcus was put into a screening chamber, where the magnetic field was reduced by a factor of 10, for 72 h. This led to a reduction (by a factor of 15) in the number and size of colonies in comparison with the control culture, which was kept in the same conditions (light, temperature, humidity), but in the natural magnetic field.

In experiments with Drosophila kept near the poles of a permanent magnet (1000 G), Levengood and Shinkle (1962) and Levengood (1965) observed an increase in the number of progeny per generation in periods of enhanced solar activity. This effect was more pronounced when the magnet was oriented with its north pole pointing east and the south pole west than when it was placed perpendicular to this direction. The effect was greater near the north pole than near the south.

Experiments with mice [Tchijevsky (Chizhevskii), 1940] kept in conditions where the strength of the magnetic field was reduced several times showed an appreciable increase in mortality in comparison with that of control animals.

A relationship between daily changes in diastolic pressure and the total leucocyte count in human blood and daily changes in the earth's magnetic field intensity was reported as long ago as 1935 (Alvarez). At about the same time the Dulls (1936) examined a large body of statistics (40,000 cases) and found a correlation between 67 magnetic storms over a period of 60 months and increases in the incidence of nervous and psychic diseases. A correlation of the same kind was established later in the period from July, 1957

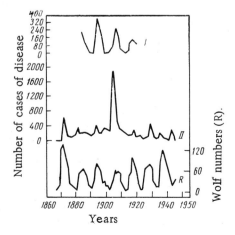

Fig. 73. Solar activity (R) and incidence of cerebrospinal meningitis in New York (I) and relapsing fever in the European USSR (II).

through May, 1961 from an analysis of more than 28,000 cases of psychic diseases for 7-, 14-, 21-, and 35-day periods corresponding to magnetic storms. It was found that the increase in the number of diseases (from admissions to psychiatric clinics) was correlated significantly (P < 0.05) with magnetic storms. Another investigation by the same authors (R. Becker, 1963, 1963b) showed a similar correlation in the case of the total number of admissions to hospitals and the number of deaths in hospitals.

Statistics for mortality from cardiovascular diseases have been compared with the times of magnetic storms (Berg, 1960). Figure 74 illustrates the correlation between the mean of 60 periods of magnetic storms and 4899 lethal cases. Recently Chernyshev (1966) discovered an effect of magnetic storms on insect activity. The observations were carried out in Turkmenia on beetles and flies (19 species) attracted to the light of an ultraviolet lamp. It was found that the number of insects attracted to the light during the night depended not only on the usually considered geophysical factors — temperature, pressure, and humidity, but also depended significantly on some other unknown external factor. On certain nights the number of insects attracted was 10-50 times greater than the expected number. Chernyshev found that the acting factor was magnetic storms. The correlation coefficient for this relationship, calculated from the data obtained for 75,336 individuals (excluding the effect of temperature and humidity), was 0.926 (P < 0.001). The effect of magnetic storms was just as pronounced as the effect of changes in air temperature. In experiments with an artificial magnetic field of intensity 1 to 1000 Oe (produced by a magnet rotating at 3 rpm) the flies showed enhanced activity if the air temperature was high (29 deg).

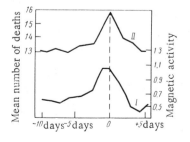

Fig. 74. Magnetic storms (I) and mortality from nervous and cardiovascular diseases (II) in Copenhagen and Frankfurt-on-Main.

11.2 The Biological Clock and Natural EmFs

The problem of the "biological clock" has long been the subject of investigations from many aspects (Bunning, 1964; Cold Spring Harbor Symposia on Quantitative Biology, Vol. 25, 1960). The main point of this problem is the evidence of the existence of a hereditarily transmitted ability of most living organisms (from unicellular organisms to man) to measure the time of day and to regulate their main physiological processes in accordance with it. In normal conditions the operation of the biological clock is correlated with periodically occurring processes in the environment — the alternation of day and night, changes in temperature and atmospheric pressure.

This periodicity is maintained for a long time even when all three or any one of these meteorological factors are constant. The period, however, may differ slightly from 24 h and the synchronism of internal and external processes is upset. The periodicity of physiological processes persists even when organisms are moved to a different longitude where all the spatially and temporally determined geophysical factors are different.

The existence of an endogenous biological clock mechanism can be regarded as established. In unicellular organisms and in all higher plants such a clock is located inside the cell. In higher animals the periodicity of physiological processes is centrally controlled, but there is also autonomous regulation in the tissues and cells. One question, however, still remains unanswered: Is the periodicity of biological processes determined only by these endogenous regulators or is it affected by periodic changes in external factors? Many scientists subscribed to the first viewpoint and believe that the periodicity of processes in organisms is determined entirely by internal mechanisms and is independent of any periodically acting factor.

The results of several investigations carried out in recent years, however, have led to a different viewpoint on daily periodicity in organisms: Advocates of this view suggest that such regulation is affected significantly by external factors which are not usually considered in experiments with so-called constant conditions, viz., the periodically varying magnetic and electric fields of the earth.

This viewpoint is held by F. Brown and his colleagues (Brown, 1959, 1962a, 1963a; Brown, 1964; Brown et al., 1956), whose results of investigations and conclusions will be discussed more fully below.

Grounds for the hypothesis of the external nature of biological rhythms were provided by the results of two series of experiments.

In the first series oysters collected on a bank at New Haven were transported in a dark tank to Evanston (16 deg further west). The oysters initially opened their valves in time with the high tides at the site of collection, but after two weeks the rhythm of opening and closing of the valves began to correspond to the ebb and flow of the atmospheric tide in Evanston. Similar effects were discovered in crabs kept in "contant conditions." In the new location the rhythm of their motor activity became adjusted to the local conditions.

The second series of experiments showed that oxygen uptake by potato plants was correlated with changes in pressure and tem-

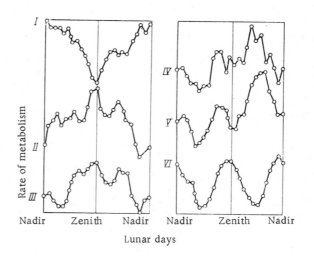

Fig. 75. Mean lunar cycle of metabolism in constant conditions. I) Rat (over 2 mth); II) mollusk (Venus) (over 8 mth); III) snail (over 1 mth); IV) alga (Fucus) (over 5 summers); V) potato (over 3 yrs); VI) crab (over 5 summers).

perature in the environment even when it was isolated from it. Such changes in metabolism had a solar-day and a lunar-day periodicity.

Further investigations showed that a similar periodicity of metabolic processes could be observed in the most diverse organisms — from algae to vertebrates — in constant conditions. Figure 75 illustrates the lunar-day periodicity of metabolism in various organisms, which is correlated with corresponding changes in barometric pressure. In most organisms (carrots, earthworms, crabs, salmanders, white mice) the metabolic cycle corresponding to the lunar days was similar to the cycle of the potato. The curves of the metabolic activity of some organisms were the mirror reflection of the curves for others.

Interesting data indicating a correlation between metabolic activity and the intensity of cosmic radiation have been obtained. The biological and physical cycles of different organisms showed either a direct or mirror-image correspondence with the change in cosmic radiation.

From all these data the authors concluded that conditions previously regarded as "constant" are not so. Being isolated from the effect of such geophysical factors as temperature, barometric pressure, and alternation of day and night, organisms continue to receive information about periodic changes in these factors, since the rhythms of physiological processes become adapted to these changes. Suspicion immediately falls on the earth's magnetic and electric fields, which can act on isolated organisms and the effect of which on cosmic-ray intensity is well known. This hypothesis

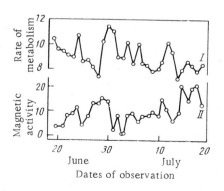

Fig. 76. Comparison of daily variations of oxidative metabolism in snails and changes in magnetic activity.

is directly confirmed by the correlation between the changes in rate of biological processes and variations in magnetic activity, as is illustrated in Fig. 76 for the rate of metabolism in snails.

A correspondence between biological rhythms and periodic changes in the earth's electric and magnetic fields can be found if the daily rhythm of some physiological processes is compared with the daily periodicity of these fields. We have made such a comparison in Fig. 77 for physiological processes in mice and man (Halberg's data, 1960) and for the change in the number of mitoses and volume of cell nuclei (Bunning, 1964).

As regards the differences between the length of the period of the endogenous biological rhythm and the actual daily rhythm Brown regards it as due to a regular phase shift between this rhythm and the external periodic process. At the same time he expresses doubt that biological rhythms differing from the daily rhythm are actually observed in nature, outside the laboratory.

We find that the most convincing views on the nature of the biological clock are those expressed by Aschoff (1960): "Organisms as open systems are always correlated to the environment. Behavior and function are results of an "inside — outside" coaction. . . . That the organism, although living in a constant environment, behaves rhythmically, we call spontaneous and endogenous. The actual value, however, of the rhythm, the frequency, is determined by all circumstantial conditions — external as well as internal."

It seems likely that the operation of the biological clock is connected with periodic electromagnetic processes within the organism and in the environment. Support for this hypothesis is found, firstly, in the fact that many periodic processes in organisms are due to the presence of electromagnetic oscillatory systems in them (see Chapter 12) and, secondly, in the above-cited results of Brown's investigation of the effect of daily periodic changes in the earth's magnetic field on corresponding periodic changes in organisms.

11.3. Orientation of Living Organisms
in Relation to the Earth's Magnetic
and Electric Fields

In a number of experiments Brown and his colleagues (Brown, 1962a, 1962b, 1962c, 1963b; Brown et al., 1960a, 1960b, 1964a,

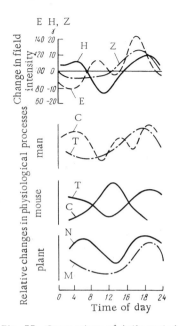

Fig. 77. Comparison of daily periodic changes in earth's electric (E) and magnetic (H, Z) fields and daily periodic rhythms in living organisms. T) Body temperature; C) eosinophil count; N) volume of cell nuclei; M) number of mitoses.

1964b; Barnwell and Brown, 1964) found that some animals installed in conditions where all the usually considered orientational factors of the environment were practically absent could still distinguish geographical directions. The experimental results obtained by these research workers indicate clearly that organisms are capable of geomagnetic orientation.

Experiments with snails, planarians, and paramecia were carried out by a method which was fundamentally the same in each case. Figure 78 shows a diagram of the apparatus. The animals were installed in a chamber on the bottom of a crystallizing dish filled with sea or river water (to a particular depth). They could leave the chamber only through a corridor oriented in a particular compass direction. The corridor opened onto an arc scale, which could be used to determine the direction of motion of the animals after leaving the corridor. The space in which the animals moved was illuminated symmetrically (on each side of the opening) or asymmetrically (so that the effect of light, especially on phototactic animals, could be assessed). A bar magnet was placed under the chamber and at different distances from it and could be oriented in any direction relative to the geomagnetic field. Each experiment entailed determination of the mean direction of motion of the animals (in groups of ten) across the arc scale. Observations of the experimental groups in particular experimental conditions alternated with observations of controls, usually by a "blind" method. The mean direction was estimated relative to the preceding or following control. In other experiments the over-all mean of the experiments was evaluated relative to the corresponding mean of all the control tests.

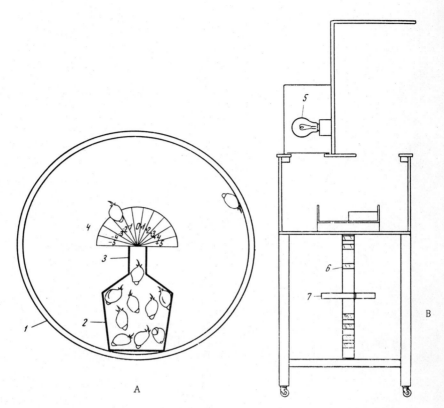

Fig. 78. Apparatus for investigation of the oriented movement of snails. A) View from above of crystallizing dish (1) with aluminum chamber (2) and corridor (3) opening onto arc scale (4): B) side view of wooden box containing lamp (5) and rack (6) for permanent magnet (7).

These experiments showed that when the animals were released in a southward direction they subsequently proceeded in a different direction according to the time of day: In the morning they tended to turn to the right, as noon approached they turned progressively to the left, and towards evening their direction of motion became similar to that in the morning. In addition to this solar daily periodicity in their direction of motion the snails also showed a lunar daily periodicity. In the presence of an artificial magnetic field of 1.5 Oe parallel to the geomagnetic field the curves representing this periodicity were of the same shape as in the absence of this field, but were displaced to some extent according to

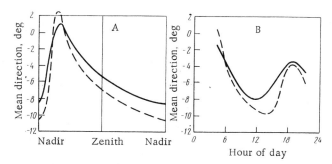

Fig. 79. Deviation of path of snails from northward direction (0°) in natural conditions (broken curves) and in presence of an artificial magnetic field of 1.5 Oe (continuous curves) in relation to time of lunar (A) and solar (B) day.

the time of day. Figure 79 shows curves illustrating the lunar and solar daily periodicity (Brown, 1963b).

The next series of investigations showed that snails could distinguish geographical directions and orientation of an artificial magnetic field. Figure 80 illustrates how the direction of the path of the snails changed when the corridor was oriented in the direction of different compass points and what changes were produced by corresponding orientations of a magnetic field (5 Oe) antiparallel to the geomagnetic field (Brown et al., 1964b). As the figure shows, the curves for the artificial field were a mirror image of the curves for the geomagnetic field.

The third series of experiments showed that the ability of snails to distinguish the direction of the geomagnetic or artificial field varied with a lunar monthly periodicity. A graph illustrating this periodicity — monthly and half-monthly — is shown in Fig. 81 (Brown et al., 1964b).

A special series of experiments was undertaken to determine the relationship between the effect of an artificial magnetic field and its intensity (Brown et al., 1964a). The greatest change in direction of the path of snails in the presence of a magnetic field opposing the geomagnetic field was manifested at field intensities close to that of the natural field (Fig. 82).

Experiments with planarians were carried out in the apparatus shown schematically in Fig. 83. A region of the scale was illumi-

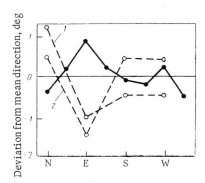

Fig. 80. Variations in direction of motion of snails in geomagnetic field with chamber corridor oriented in different compass directions (broken lines) and same orientations of an artificial magnetic field of 5 oe (solid curve). 1) Emergence of snails into symmetrically illuminated field; 2) emergence into asymmetrically illuminated field.

nated through an opening in the top of the darkened chamber. To avoid the horizontal light beam* directed along the polar axis from the right, the planarians swam to the center of the arc scale. The angle at which the animals crossed the scale was then recorded (Brown, 1962a, 1962b, 1963b; Brown et al., 1960).

The first series of experiments showed that planarians could distinguish compass directions and the orientation of an artificial magnetic field of 10 Oe. As Fig. 84 shows, planarians deviated to the right (in a clockwise direction) from the polar axis when it pointed north and south and to the left when it pointed west and east. In experiments with the artificial magnetic field oriented in different directions the deviations were of the opposite nature.

The second series of experiments showed a lunar monthly periodicity in the deviations of planarians from the northward direction. In the autumn and winter months the greatest deviations from the mean directions were repeated every new moon, while in the spring months the period of repetition was half as long.

The third series of experiments showed deviations of the path of planarians in the geomagnetic and north- or east-directed artificial magnetic fields of 4 Oe to 0.25 Oe, when the polar axis of the scale pointed toward the north. Unlike for fields of 10 Oe the sign of the compass response to these weaker fields was the same as that to the natural field. A change in sign of response clearly occurs between 5 and 10 Oe.

_____

* Planarians show negative phototaxis.

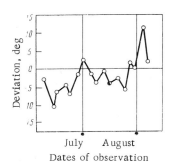

July   August
Dates of observation

Fig. 81. Lunar monthly periodicity in difference between directions of motion of snails on emergence to the east and north (black circle indicates new moon).

A monthly rhythm in the nature and degree of deviations was also discovered. At new moon the deviations were 10° left of the axis and at full moon they were 10° to the right. This rhythm varied in the course of the year.*

The apparatus used for experiments with paramecia consisted of a petri dish with a chamber and emergence corridor. The arc scale was in the field of view of a stereoscopic microscope (made mainly from nonmagnetic materials) and was weakly illuminated from below (through a water filter, which absorbs heat rays). The whole apparatus was in a dark room. The experimental procedure consisted in observation of the paths of the paramecia across the arc scale (Brown et al., 1962b).

The paths of the paramecia in the geomagnetic field with the chamber corridor oriented to the north were compared with the paths observed in the presence of an additional east- or west-oriented artificial field of 1.3 Oe. The results of such investigations are shown in Fig. 85. It is obvious that the distribution of directions of motion in the two cases is the same (curves in Fig. 85B). A correlation between changes in the direction of motion of the para-

---

*Editor's note: Other investigations with planarians revealed that the receptive system for weak magnetic fields exhibits physiological adaptation to slightly altered magnetic field strength, thereby displaying a property common among more conventional receptive systems (Brown and Park, 1965). The planarians, in another investigation, were demonstrated to be able to associate the direction of an experimental light source with direction of the horizontal vector of a magnetic field, whether the the natural one or an experimentally imposed one and retain this information for at least many minutes (Brown and Park, 1967).

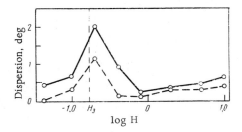

Fig. 82. Mean deviation (dispersion) of direction of motion of snails due to artificial magnetic field (opposed to geomagnetic field) of
different intensity from directions of motion in
geomagnetic field alone. The broken curve
represents the mean values for 6 mth, the continuous curve for 3 mth; H_E is the geomagnetic
field strength.

mecia (Fig. 85C) and the phases of the moon was detected: The
maximum number of clockwise turns (from the polar axis of the
scale) was observed four days after new moon, and the maximum
number of anticlockwise turns was observed four days after full
moon. Deviations four days after new moon were correlated with
the lunar phases with a coefficient $r = 0.76 \pm 0.09$.

The protozoa *Volvox aureus* can orient itself relative to the
geomagnetic field and is sensitive to weak magnetic fields, as Palmer (1962, 1963) found by the same method as in the experiments
with paramecia (the only difference was that the Volvox were induced to leave the chamber by a brief illumination). Three series
of experiments were carried out: I) in the geomagnetic field (control); II) in an additional field of 5 Oe parallel to the geomagnetic
field; III) in an additional field of 5 Oe perpendicular to the geomagnetic field.

Daily observations over many months showed the following
relations between the number of clockwise turns of Volvox: in the
second series of experiments there were 43% more than in the
first; in the third there were 150% more than in the first and 75%
more than in the second.

Interesting results were obtained in an investigation of the
geomagnetic orientation of snails and planarians in an artificial

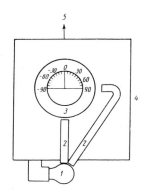

Fig. 83. Apparatus for orientation experiments with planarians. 1) Opalescent lamp; 2) light guides; 3) petri dish with arc scale; 4) darkened chamber; 5) direction of zero axis of scale.

electrostatic field between capacitor plates (Brown, 1962a, 1962c). It was found that the directions of motion of these animals in a field with a strength of only 2 V/cm perpendicular to the body axis differed from the directions of motion observed in the geomagnetic field alone. Figure 86 shows how the different polarity of the electrostatic field altered the direction of motion when the chamber corridor was parallel or perpendicular to the geomagnetic field. More thorough investigations (Webb et al., 1961) suggested that snails can distinguish the direction of the lines of an electrostatic field (positive plate of capacitor right or left of the body axis). Similar results were obtained in experiments with planarians with the emergence corridor oriented in different compass directions.

We recall again the change in the speed of planarians swimming in a horizontal glass tube placed in a vertically directed electrostatic field with a strength of 15 V/cm (Kenneth Penhale, cited by Brown, 1962c), the diurnal periodic reaction of marine snails to a vertical electrostatic field of 15–45 V/cm (Webb et al., 1959) and, finally, the sensitivity of living organisms to electrostatic fields, as described in the second part of this book.

We have described only the main results of numerous investigations (thousands of experiments, tens of thousands of observations) carried out by Brown and his colleagues over several years. These data were obtained by proper experimental techniques and careful statistical analysis of the observations. Hence, the reliability of the experimental results is beyond question. The main hypotheses expressed by the authors on the basis of their experimental results are sound. These hypotheses reduce mainly to the following:

1. Living organisms have mechanisms which act as a "biological compass" and a "biological clock." These mechanisms enable organisms to orient themselves relative to the earth's electric and magnetic fields in accordance with periodic changes in geophysical factors of a worldwide or local nature.

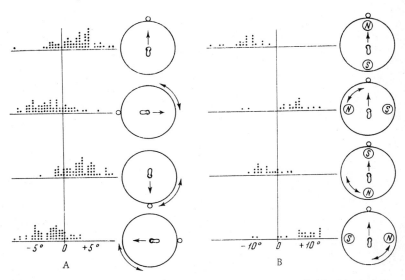

Fig. 84. Directions of motion of planarians in geomagnetic field with chamber corridor directed toward different compass points (A) and in an artificial magnetic field of 10 Oe with corresponding orientations of its north pole (B).

2. These mechanisms are "tuned" to weak natural fields and their operation is probably based on trigger-type reactions, the nature of which depends on other geophysical factors. In investigations of the biological action of magnetic and electric fields of the high strength usually used in laboratory experiments these reactions may be inhibited or totally suppressed.

3. These mechanisms are adaptive in relation to environmental changes and, of course, within the organism they act very closely with the main systems responsible for coordination of the vital processes of organisms in their interaction with the environment.

It has been established directly in some investigations that insects can orient themselves in space relative to the earth's magnetic field. According to G. Becker (1963a, 1963b), cockchafers, bees, crickets, wingless termites, and many flies are capable of such orientation. For instance, in calm weather or when the wind is light flies almost always land in an east— west or north— south direction, irrespective of the position of the sun. Resting flies strive to maintain this direction of the body or alter it by a rapid

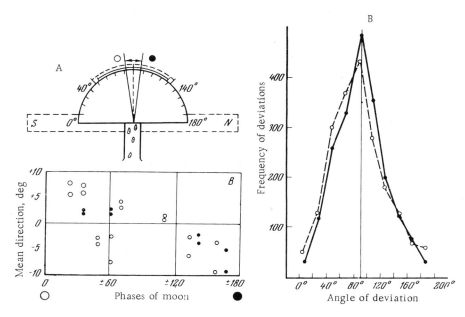

Fig. 85. A) Orientation apparatus for paramecia; the spread of the directions of motion of the paramecia at full moon (O) and new moon (●) is shown on the arc scale; B) frequency of deviations in direction of motion of paramecia through different angles from scale axis (90 deg) in geomagnetic field (broken curve) and artificial magnetic field of 1.3 Oe oriented in an east−west direction (continuous curve); C) deviations in relation to phases of moon.

movement through 90°. In the field of a permanent magnet 100 times stronger than the geomagnetic field, insects become excited, but after some time they orient themselves parallel or perpendicular to the magnetic lines of force. An electrostatic field with a strength 100 times greater than the strength of the earth's electric field renders flies inactive. Very strong magnetic fields also suppress the activity of insects.*

---

* Editor's note: Recent work has shown that the orientation of the bee dance on a vertical honeycomb is affected by very small experimentally induced alterations in the ambient magnetic field (Lindauer and Martin, 1968). The observations indicated that the bee's orientation is affected by both small increases or decreases in field strength, suggesting that normally there exists a nice adjustment of the bees to geomagnetism.

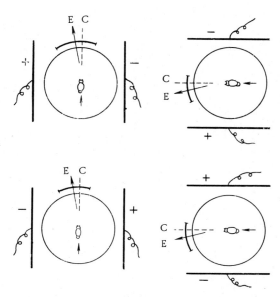

Fig. 86. Illustration of effect of electrostatic field
(2 V/cm) on motion of snails. E) Experiments in electric field; C) control (geomagnetic field alone).

Several investigations have shown that the growth of plants
and their spatial orientation depend on the geomagnetic field.

It has been found (Krylov and Tarakanova, 1960, 1961; Koryak, 1966) that seeds of cereals (corn, wheat) sown with the radicle
pointing south germinate more rapidly (1-3 days earlier) than those
sown with the radicle pointing north. Chubaev's experiments (1966)
showed that growth stimulators enhance this effect, while inhibitors
not only reduce the effect, but even cause the reverse effect. A
similar reversal of the effect is produced by an artificial magnetic
field of 0.4 to 20 Oe. The effect also depends on the phases of the
moon: When the seeds are sown at new moon the effect is least
distinct.

Geomagnetotropism is manifested in the bending of seedlings
to the south if the seeds are sown with the radicle pointing north.
The photograph shown in Fig. 87 (from Krylov and Tarakanova's
paper, 1961) illustrates this effect and also the difference in rate
of germination of seeds in relation to their orientation relative to

the earth's magnetic field. Abros'kin (1966) observed geomagnetotropism in the northern white cedar: The best developed skeletal
branches given off from this tree in the east—west direction subsequently grew in approximately vertical planes running mainly in
the north—south direction; this type of orientation of the branches
was observed in 136 out of 152 trees.

Finally, we will mention the experiments of Pittman (1962) in
Canada, who showed that wheat seeds sown with their axis of symmetry in a north—south direction gave a better crop than those
sown at random.

Thus, the discussed experimental results indicate quite clearly that animal organisms have "orientational systems" which re-

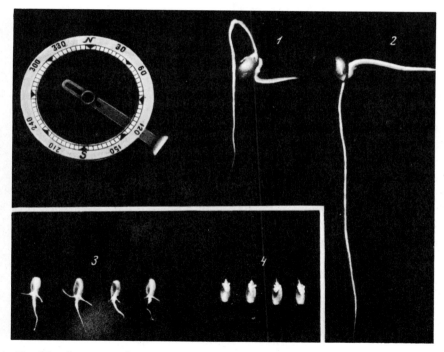

Fig. 87. Illustration of geomagnetotropism in the germination of corn and wheat
seeds. Corn seeds with their radicle pointing to the south (2) germinated more rapidly than those with their radicle pointing north (1), and in the latter case the
seedling turned toward the south; 3, 4) similar comparison of wheat seeds.

spond to the earth's magnetic and electric fields. This again gave rise to the hope of verifying the hypothesis, put forward more than a hundred years ago by the Russian academician Middendorf, that birds can orient themselves relative to the earth's magnetic field.

In the past 20 years there have been several attempts to verify this hypothesis experimentally, but they have not given definite results so far.

Yeagley's experiments (1947, 1951) appeared quite convincing. Proceeding from the assumption that orientation is effected by means of the geomagnetic field he carried out the following experiments. Pigeons trained to return from great distances to a portable dovecote were taken thousands of miles away to a locality where the indices of the geophysical factors were the same as at the training site. In this case the pigeons easily found the long way back to the dovecote. In other experiments Yeagley found that if small magnets were attached to the pigeons' wings they lost their ability to navigate.

Several repetitions of Yeagley's experiments (D. Gordon, 1948; Van Riper and Kalmbach, 1962; Griffin, 1955; and others), however, did not confirm has results. Attempts to produce a conditioned reflex to a magnetic field in pigeons were a failure, although a weak effect of the field on their behavior was observed in maze-learning experiments (Neville, 1955). For these reasons many scientists are sceptical of the geomagnetic hypothesis of bird navigation and point out that from the evolutionary viewpoint such navigational ability should be observed in other animals too. This, however, cannot be regarded as a convincing argument. As we have seen, orientation relative to the earth's magnetic field has been experimentally discovered in various species of organism.* In addition, recent observations indicate the possibility of geomagnetic orientation (as well as orientation relative to an electric field) in fish too.

---

* Editor's note: The orientation of birds during their period of migratory unrest, deprived of all obvious geographic directional cues, has recently been shown to be altered by experimental manipulations of their magnetic-field direction at strengths close to those of the earth's natural field (Wiltschko and Merkel, 1966; Wiltschko, 1968). The receptive mechanism which is concerned appears to be highly adapted to the natural ambient field strengths.

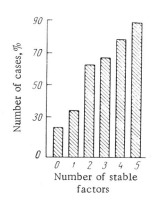

Fig. 88. Frequency of cases of movement of fish along magnetic meridian in relation to number of stable environmental factors (temperature, electrical conductivity and transparency of water, presence of current, depth of location of fish.

Poddubnyi (1965) found that 87.5% of fish transplanted to a new basin and 50% of the local individuals moved in the magnetic meridian. This occurred in particular portions of the migration route, where the fish were in water with uniform hydrophysical indices, and was manifested more often the greater the number of stable indices. This is illustrated by the diagram shown in Fig. 88. The author believes that the obtained results indicate that fish are capable of geomagnetic orientation. This possibility is confirmed by the above-described experiments (see Section 7.2) in which conditioned reflexes to a magnetic field were elaborated in fish.

We are inclined to subscribe to the opinion of scientists (Dement'ev, 1965; Manteifel' et al., 1965) who believe that the problem of geomagnetic orientation in birds and fish merits investigation. A fertile concept in this connection is the above-mentioned hypothesis of Brown that the navigational systems of living organisms area combination of a biological "compass" and a "clock." The experimental results discussed in this section indicate the necessity of a general biological approach to the investigation of the geomagnetic and geoelectric orientation of living organisms. The fact that such orientation and the presence of a "biological clock" have been observed even in the most primitive unicellular organisms suggests that the simplest systems responsible for these vitally important abilities came into existence at the dawn of evolution and have subsequently become more complex.

11.4. Effect of Natural Low- and High-Frequency EmFs on Chemical and Biological Systems

So far we have considered investigations directed toward the discovery of a relationship between various biological phenomena and the earth's magnetic or electric fields, which vary slowly with a many-year, annual, monthly, or daily periodicity. However, as already mentioned (Section 2.3), there is a broad spectrum of EmFs, produced by atmospheric discharges in the natural external environment. The total intensity of these EmFs over the whole planet also varies periodically — in time with the 11-year cycle of solar activity, in the course of the year, and during the day. This naturally gives rise to the question: Do these EmFs play any significant role in the vital activity of organisms or should they be regarded only as interference to which organisms have become adapted in the course of evolutionary development? As regards investigations on animals, very little experimental information on this question has been obtained so far. There are only experiments on hymenoptera (Maw, 1961), which showed that the rate of egg-laying increased when the insects were screened from fluctuating natural EmFs in the presence or absence of an artificial electrostatic field of 1.2 V/cm (which is close to the strength of the earth's electric field). The same effect was observed in the case where insects, screened from natural EmFs, were exposed to the action of artificial EmFs (0.8 V/cm) fluctuating like natural fields.

Over a period of 15 years Prof. Georgio Piccardi in Florence conducted systematic investigations which showed very definitely that low- and high-frequency natural EmFs affect the course of chemical reactions and various biological processes in vitro. Within a year after the start of these investigations in Italy they were taken up in several countries on all continents (in 15 countries so far). We will discuss the main methods and results of these investigations from Piccardi's monograph and a special collection under his editorship (Relations entre phenomènes solaires . . . , 1960; Piccardi, 1962, 1965), where a large bibliography was given.

Working with colloidal solutions Piccardi found that the same precipitation reaction occurred at different rates at different times, although other conditions were the same. This irreproducibility could not be attributed to experimental errors and Piccardi suggested that it was due to the action of spatial forces, most probably of an electromagnetic nature, on water.

The grounds for this hypothesis were the results of a comparison of the fluctuations in the precipitation rate in the case of hydrolysis of bismuth chloride in ordinary and screened test tubes.

Piccardi first verified the regular nature of the fluctuations in the precipitation rate when this reaction occurred in ordinary test tubes by comparing two randomly chosen groups of ten tubes. As was to be expected, the distribution of the frequencies of cases in a particular percentage difference in the precipitation rate was Gaussian (with an average difference in rate about 50%). However, a comparison of the precipitation rate in similar groups of unscreened and screened tubes showed that the distribution was asymmetric (with an average difference of 70%), which indicated the effect of external forces of an electromagnetic nature.

Piccardi used this comparison of the precipitation rate in screened and unscreened tubes as the main test for further investigations and determined (as a percentage) the frequency of cases of difference in precipitation rate (T). He also used a modification of this test in which the solutions were made up not with ordinary water, but water "activated" by electromagnetic fields in a wide frequency range (10 Hz to 4 kHz) (these fields were produced by the sliding of a drop of mercury in a glass sphere submerged in the water). In the main experiments with bismuth chloride Piccardi used three types of tests in which the precipitation rate was compared in the following two conditions:

	Condition I	Condition II
P-test	Normal water Unscreened	Normal water Screened
F-test	Normal water Unscreened	Activated water Unscreened
D-test	Normal water Screened	Activated water Screened

The effects of various geophysical factors (temperature, atmospheric pressure, light cosmic radiation, etc.) were not excluded in any of these tests. These tests were first used in a series of preliminary investigations to determine the difference in the precipitation rate of bismuth chloride in relation to the experimental conditions, and then in the series of main investigations of the relationship between this effect and changes in external factors in time. Table 16 gives the results of the preliminary investigations and some of the features discovered in this case.

Table 16. Relation between Percentage Difference (T) in Precipitation Rate of Bismuth Chloride and Experimental Conditions

Investigated relationship	Type of test	Description of compared experimental conditions	Extimate of T, %	Results
On method of screening	D	In open Under copper plate Inside copper chamber	54.5 37.0 37.5	Each method of screening equally effective. Forces apparently act in downward direction
On screen material	D	In open (0 $\Omega^{-1} \cdot cm^{-1}$) Iron ($10 \cdot 10^4 \Omega^{-1} \cdot cm^{-1}$) Aluminum ($35 \cdot 10^4 \Omega^{-1} \cdot cm^{-1}$) Copper ($58 \cdot 10^4 \Omega^{-1} \cdot cm^{-1}$	46.4 45.4 42.2 39.6	Effect depends on conductivity of screening material; this is characteristic of effectiveness of screening from EmFs
On structure of screen	P	In open — aluminum plate In open — plate covered with aluminum paint Aluminum plate — plate coated with aluminum paint	76.8 72.6 45.6	Effect of screening by solid metal different from that of aluminum paint (isolated metal particles). Plate provides screen against EmF of 10 kHz but paint does not. Further investigation required.
On screen thickness	D & P	0.3 mm (lead) 15 mm (lead) 30 mm (lead)	39.7-40.4* 42.9 42.9-41.4*	No significant dependence on thickness of screen, which was varied by a factor of 100
On thickness of thin screening coatings	P	In open — gold film (0.0005 mm) on cardboard In open — gold evaporated onto cellophane (50 Å) Gold on cellophane — gold film	71.0 45.0 76.0	Gold film a more effective screen than gold evaporated onto cellophane
On height above sea level	D	800 m — at minimum of solar activity 2000 m — at minimum of solar activity 328 m — at maximum of solar activity 3578 m — at maximum of solar activity	51.9 54.7 45.4 54.3	Effect depends on height above sea level; more pronounced at maximum of solar activity

Investigated relationship	Type of test	Description of compared experimental conditions	Estimate of T, %	Results
On height above sea level	F	800 — at minimum of solar activity	48.6	
		2000 m — at minimum of solar activity	55.0	
		328 m — at maximum of solar activity	38.3	
		3578 m — at maximum of solar activity	47.3	
	P	328 m — at maximum of solar activity	64.0	
		3578 m — at maximum of solar activity	43.2	
On time of day	F D P	Midday — midnight (experiments in Tubingen, F. R. G. from May through July 1957)	37.6-40.4 54.7-58.1 36.2-34.0	No appreciable difference in effects at midday and midnight. Experiments conducted in Brussels (August-September 1958) showed difference for P test of 56.3-47.1. Additional investigations necessary.

*The first figures are from experiments carried out between July 1953 and April 1954
and the second from experiments carried out between January and April 1954.

The main investigations, carried out between March, 1951 and October 1960, included more than 250,000 tests. They showed that the effect depended on the solar activity. This was revealed in the secular, annual, and short-period variations of the effect.

The secular variations are illustrated in Fig. 89, which shows that the minimum for the D and F tests (from the annual values) was observed in 1954 — the period of minimum solar activity.

The annual variation is shown in Fig. 90. This figure clearly illustrates a periodicity for the D test with minima in March of each year. It is characteristic that the same feature was observed much later — in 1961-1962; the minima in August were not significant. F-test experiments in years of maximum solar activity

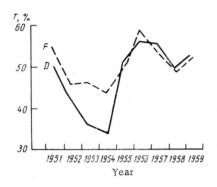

Fig. 89. Comparison of results of D- and F-test experiments over a period of nine years. The minimum T in 1954 coincides with the minimum of solar activity.

showed approximately the same variations. In years of low solar activity the periodicity was the same, but the minima were observed in summer and the maxima in winter.

Short-period variations were manifested in a correlation between the effect evaluated by the D test and cosmic-ray intensity, as Fig. 91 shows. The variations in the F-test experiments coincided precisely with solar flares and magnetic storms. Investigations of the P-test variations also showed a correlation with solar activity — positive when the screening chamber was used and negative when the plate was used.

In 1960 series of experiments with a new test — polymerization of acrylonitrile (A-test) — were begun. The effect was assessed by comparing the total weight of the acrylonitrile in a blackened cardboard chamber and in a copper screening chamber (ten tubes chamber). These experiments showed that the percentage difference in the weight of the unscreened and screened polymer varied from month to month and was correlated positively with corresponding variations in P-test experiments. This means that the factors affecting polymerization and precipitation of bismuth chloride are presumably the same. Since polymerization is slow, these factors must evidently act over a long period.

The described experimental results led Piccardi to postulate that one of the causes of the observed effects is the influence of EmFs of atmospherics. Experiments with artificial EmFs confirmed this hypothesis. In these experiments the precipitation rates of bismuth chloride were compared when the tubes were at distances

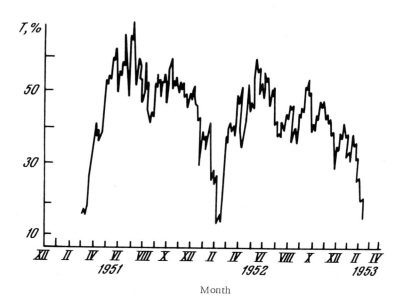

Fig. 90. Results of D-test experiments in southern hemisphere during two-year
period

of 2 m from an artificial EmF source with a frequency of 10 or 120
kHz. In other words, the effect of reducing the intensity of the acting EmFs by a factor of 100, which is equivalent to screening of
the tubes, was investigated.

It was found that at a distance of 2 m from the source the
precipitation was much slower than at a distance of 20 m and the
variations of this difference from day to day were the same as the
corresponding variations in the P-test experiments (Fig. 92A).
Such a correlation was traced over several months, as Fig. 92B
illustrates.

The effect of natural EmFs with a frequency of 10 kHz was
investigated in polymerization experiments. Such an effect was
found and its variations were correlated with corresponding variation in the P-test experiments. The experimental results, however
indicated the influence of certain external forces, even when screening was used.

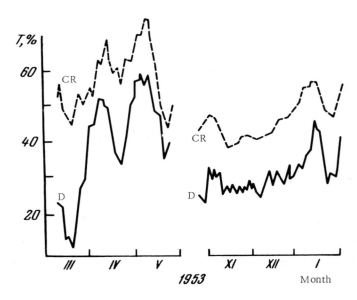

Fig. 91. Comparison of results obtained in D-test experiments with
changes in cosmic-ray intensity (CR).

Research over many years led Piccardi to the conclusion
that the observed effects were due to the effect of EmFs of atmospherics and to some even stronger factors on the water in which
the chemical processes took place. If this is the case, we can expect these factors to affect biological processes. This hypothesis
is confirmed by experiments with various biological tests. For
instance, clotting of human blood in screened (with 0.1 mm of copper) capsules was slower than in unscreened, and this difference
was correlated with the corresponding difference in the chemical
P-test experiments (precipitation of bismuth chloride). A similar
effect was observed also in the reduction of the rabbit erythrocyte
sedimentation rate. An effect of screening on the growth of bacterial cultures (staphylococci and colon bacilli) has been found.

On analyzing the results of his numerous investigations Piccardi came to the general conclusion that living organisms and the
medium surrounding them are subject to the influence of various
cosmic factors. In view of the similar nature of the observed reac-

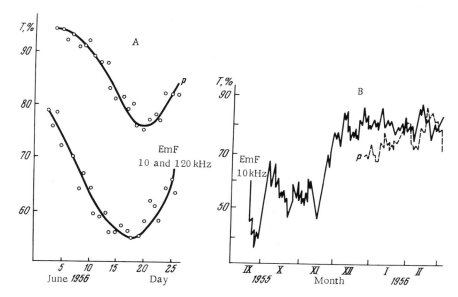

Fig. 92. Comparison of effect of EmFs of different intensity on precipitation rate of bismuth chloride (test tube at distances of 20 and 2 m from EmF source) with results obtained in P-test experiments. A) Over 20 days; B) over half a year.

tions of inorganic and biological colloids to cosmic factors he suggested that the reacting component in living organisms is a "non-living substrate" — water and colloids.

We will discuss the validity of this hypothesis later. Here we will merely cite Piccardi's conclusions which have a direct or indirect bearing on the problem of the biological effect of natural EmFs.

1. The observed irreproducibility — fluctuations of various processes in living systems and inorganic colloid systems — cannot be attributed either to experimental errors or to the effect of the usually considered geophysical factors (temperature, pressure, humidity, etc.). These fluctuations are due to the biotropic effect of cosmic forces of electromagnetic or corpuscular nature — electromagnetic radiations of cosmic or terrestrial origin, variations in the earth's electric and magnetic fields, solar activity with all its manifestations and, possibly, other as yet undiscovered cosmic factors.

2. The periodicity, or cyclicity, of fluctuational phenomena is due to the spiral motion of the earth in the galaxy. This motion accounts for the 27-day, 135-day, and other periodic variations observed in experiments with chemical and biological tests, the correlation between these variations and solar activity, the dependence of these variations on the geographical latitude, and so on.

3. Fluctuational phenomena in heterogeneous, nonequilibrium, thermodynamically open systems (biological and inorganic) are not due to processes occurring at the atomic and molecular levels, but to processes at the level of larger and more complex heterogeneous structures, where external factors of very low energy can cause considerable effects.

4. The proper approach to the investigation of the mechanisms of the action of external factors on such systems is the consideration of the structural characteristics of water, solutions, and aqueous colloidal systems. The possibilities of geometric and energetic variations in such systems are quite unlimited. The investigation of the mechanism of the effect of electromagnetic radiations and variations of electric and magnetic fields on these systems will necessitate the determination of the sensitivity of individual elements of the systems of these factors.

5. Effects of electromagnetic and corpuscular nature must be distinguished. Particles act chaotically and sporadically and only on some parts of the organism, whereas the action of EmFs affects the whole organism — it is "total" in the sense that it acts in any circumstances.

The experimental work and theoretical arguments of Piccardi were further developed in the work of other Italian scientists (Maletto and Valfre, 1966; Masoero et al., 1966), who regard environmental EmFs as ecological factors which have a significant effect on the vital activity of organisms — on phenotype formation, on reproductive processes, on regulation of the size of populations, on the assimilation of nutrients by organisms, and so on. This effect can be direct — on the entire organism, on organs, tissues, cells, or macromolecules, or it can be indirect — by action on plants and water consumed by animals.

Numerous investigations (Aghina et al., 1965; Caramello et al., 1965; Cellino Tosi et al., 1965; Maletto et al., 1965a, 1965b,

1965c, 1966a, 1966b; Masoero et al., 1965, 1966; and others) have given results which support this concept to some extent:

1. A statistical analysis of 89,712 cases of fertilization of cows showed that fecundity was positively or negatively correlated with changes in solar activity (expressed in Wolf numbers). The authors suggest that low-frequency environmental EmFs play the main role in this effect.

2. Screening of plants usually used to feed farm animals (with other environmental factors constant) led to a significant reduction in the rate of nitrogen metabolism. The authors regarded this effect as due to a diminution of the effect of low-frequency EmFs on plants.

3. Exposure of guinea pigs to static and low-frequency fields in the laboratory led to progressive erythrocytopenia, partial leucopenia, and an increase in hemoglobin content. This was accompanied by the appearance of immature erythrocytes, which, as exposure continued, became more and more abundant.

4. The udders of cows were exposed twice a day (after the morning and the evening milking) to a slowly varying magnetic field produced by a rotating permanent magnet. Several such exposures led to significant changes in the fat content of the milk due to an extension or reduction of the carbon chains of the fatty acids.

5. Animals (mice, guinea pigs, chickens) given water "activated" by low-frequency fields showed a lower rate of weight increase during growth, enhanced water retention in some organs (heart, spleen, white muscles), and an increased amount of unsaturated fatty acids in the fat deposits.

From an analysis of these experimental results and the numerous experimental data obtained by other researchers Maletto and Valre (1966) came to conclusions in agreement with those of Piccardi. They think that the effect of natural and artificial EmFs not only on water and plants, but also on entire organisms, is due to the direct action of these fields on biochemical processes, and primarily on the structure of water in which these processes take place.

We think that such a mechanism is probable, but it is not the only possible one, especially in the case of the reaction of complex

organisms to EmFs. Direct effects of EmFs at the molecular le-
val can be regarded as the main (though not the only) way in which
these fields affect the vital activity of unicellular organisms, plants
and, finally, primitive multicellular organisms (which usually live
in an aqueous medium). However, even in those organisms, and
particularly in complex organisms with a highly developed centra-
lized control system, the changes in vital processes in relation to
changes in external EmFs cannot be attributed entirely to direct
effects at the molecular level. It is more likely that EmFs act on
some of the macroscopic systems which comprise the complex
scheme of hierarchic regulation and interconnections in the organism.*

11.5. Mechanisms of the Effect
of Natural EmFs on the Vital
Activity of Organisms

As we have seen, natural environmental EmFs either have a
regulating influence on living organisms, thus assisting the normal
course of vital processes and the normal interaction of organisms
with the environment, or they have a disturbing effect, leading to
some abnormlity of these processes and interactions. Periodically
(or cyclically) varying EmFs have a regulating effect. This effect
is manifested in the synchronization of biological rhythms and in
the spatial orientation of organisms. Sporadic changes in EmFs
disturb vital processes to some extent and this is particularly appreciable during development and in the pathological states.

Possible mechanisms of this kind of action of natural EmFs
can be deduced from the experimental information discussed in
this chapter and from comparisons with the results of investigations
of the biological effects of artificial EmFs in the corresponding
frequency ranges (Presman, 1965a, 1965b, 1967a, 1967b, 1967c).

In examining the main features of the reactions of entire organisms of various species to artificially produced EmFs we en-

---

* Editor's note: Very weak, 10 Hz fields have been demonstrated to influence
the period of the circadian rhythm of man held in constancy of illumination and temperature (Wever, 1967, 1968).

countered a significant difference in the nature of the d i s t u r -
b a n c e s   i n   t h e   r e g u l a t i o n   of physiological processes due to
the action of EmFs on the central and peripheral systems. The
peripheral systems react rapidly to EmFs with any parameters,
whereas the central systems react appreciable only to EmFs of
low intensities and the latent period is long.

   The situation is different when we consider the question of
the r e g u l a t i n g i n f l u e n c e of natural EmFs. Organisms must
have become adapted to their influence in the course of evolution
and, hence, must have systems which react selectively only to
EmFs which carry useful information and are protected from sporadic changes in EmFs. Incidentally, in the light of Salzberg's
above-mentioned definition the latter effects are noise rather than
interference. It is reasonable to assume that useful information is
perceived by the central systems, which coordinate vital processes
with natural changes in the environment. As already pointed out,
natural EmFs are the most suitable carrier of such information.
If this is so, the central systems must be receptive in a very narrow band of frequency and amplitude and, consequently, must be
slow in response. Hence, they will be capable of storing electromagnetic information. The peripheral systems, which have become adapted in the course of evolution to the perception of all
external influences (otherwise the organisms could not have existed)
and can "filter out" useful from harmful signals, must have opposite properties: fast response and broadbandedness in both frequency and amplitude.

   If we accept this hypothesis we inevitably come to the conclusion that such a differentiation appeared at the first stages of evolution — in unicellular organisms. These organisms have been
found to possess a "biological clock," which coordinates the rhythm of biological processes with periodic changes in the environment. These organisms can orient themselves relative to the
earth's magnetic and electric fields. The vital functions of these
organisms are disturbed both by sporadic changes in natural EmFs
and by artificial reduction of the intensity of the periodic action of
these fields. It is true that the peripheral excitable structure of
unicellular organism protects them very ineffectively from the
harmful effect of EmFs. Not only a magnetic field, but low-frequency EmFs of atmospherics, can penetrate the cells and reach

222

its "central systems" — the nucleus and organelles. On the other hand, the aqueous medium in which most unicellular organisms live protects them from the action of higher-frequency EmFs. Finally, active protection may be provided by the difference in the rate of response of the systems: The peripheral system, which responds rapidly, can "warn" the slowly reacting central system of harmful action and, hence, give the latter the opportunity to protect itself by "internal means" — reduction of the sensitivity to EmFs.

In multicellular organisms the selective response and protection of the central systems become more and more advanced and complex as the tissues become progressively more differentiated and the nervous system develops. Although (as many scientists believe) autonomous intracellular regulation of the main biological rhythms is still a feature of complex organisms, up to higher mammals, there is also a multistage central regulation of the rhythmicity of physiological processes, which is coordinated with periodically varying external factors. Information can be conveyed from the environment to such organisms not only by the periodically varying electric and magnetic fields of the earth. It can apparently also be conveyed by the EmFs of atmospherics and the radio emission of the sun, which vary periodically in intensity. At the same time, the central control systems of complex organisms are reliably protected from interference due to sporadically varying EmFs of all frequency ranges.

There is still not sufficient experimental or theoretical basis for definite conclusions regarding the fundamental nature of the biological systems which receive EmF information from the environment and protect the organism from electromagnetic interference. In this connection we must confine ourselves to general considerations. Thus, it can be postulated that the high sensitivity of organisms to weak environmental EmFs may be based on mechanisms of spatial and temporal summation, where signals are received simultaneously by n elements or signals repeated n times are received. As we mentioned in the introduction, in these cases the signal-to-noise (or interference) ratio is increased by a factor $\sqrt{n}$.

Such a mechanism could ensure synchronization of the biological clock with the daily periodic changes in environmental EmFs, which can be regarded as a "differential" Zeitgeber. Another "pulsed" Zeitgeber of biological rhythms could be the repetition of weak magnetic storms at 27-day intervals. Finally, the 11-year and seasonal cyclicity in the behavior and functioning of organisms may be regulated by corresponding changes in the level of magnetic activity — a "proportional Zeitgeber."

In discussing the correlation between the distubances of physiological processes and increases in the intensity of EmFs in periods of solar flares we must consider the fact that the intensity of EmFs of one type increases practically simultaneously with solar flares and after a considerable delay in the case of increases in intensity of EmFs of other types: In the case of "bursts" of radio emission of the sun and the sudden increases in atmospherics the effect is simultaneous (9.5 min after the burst or flare), while in the case of magnetic storms it is delayed (up to 26 h after the flare).

We have cited evidence of the direct effect of artificial and natural EmFs on protein solutions, on colloid systems, and finally, on water. How can this affect the vital functions of organisms? As regards unicellular organisms, they can react directly to changes in the aqueous medium and, in addition, EmFs may cause certain changes in the protoplasm itself. Although periodic and sporadic changes in natural EmFs may act directly on the liquid media of complex organisms, they can hardly lead to an appreciable disturbance of homeostasis. It is quite possible, however, that such factors may have an indirect effect on complex organisms in cases where the latter consume water, or unicellular organisms which have previously been "activated" or "disorganized" by the action of EmFs. We have already mentioned that animals which were given water "activated" by an EmF with a frequency of 10 kHz put on less weight than controls (which received ordinary water) and that similar changes were observed also in their progeny, which were given ordinary water. It is characteristic that in both cases the effect was particularly pronounced in periods of solar activity (Valfre et al., 1964).

In conclusion we mention some hypotheses on the effect of environmental EmFs on the rhythm of brain biopotentials.

It was recently found that the earth's magnetic field is subject to micropulsations with frequencies of 0.01 to hundreds of

Hertz. R. Becker (1963b) drew attention to the fact that these pulsations are most pronounced in the range 8–16 Hz and suggested that there might be a connection between the alpha rhythm of the electrical activity of the brain and these micropulsations. Herron (1965) discovered another distinct range of micropulsation frequencies — from 0.029 to 0.031 Hz, which corresponds to the ultraslow fluctuations of the brain potentials. Of course, comparisons of such kind cannot in themselves provide a basis for a definite conclusion regarding the connection between these two phenomena, but it is impossible a priori to rule out such a possibility in view of the correspondence between the total frequency range of the magnetic field and the total frequency range of the variations in biopotentials (from the ultraslow variations in the brain to the frequency spectrum of impulses in the nerves).

As regards the possibility of the influence of external EmFs on the alpha rhythm, we must mention the results of experiments with people exposed to EmFs with a frequency close to 10 Hz (Wiener, 1963). The subjects experienced an unpleasant sensation similar to that previously produced by exposure to pulsating light with a frequency of 10 Hz. Vibrations of the same frequency predominated in the alpha rhythm in the case of both kinds of treatment.

Thus, it seems probable that the regulating and disturbing effects of EmFs on living organisms are due to two causes: First, these fields act directly on biochemical processes occurring in the organism, and second, they act on the peripheral and central systems of the organism which are responsible for regulating vital processes in correspondence with environmental changes. While the effect of EmFs on the simplest organism probably involves both causes, the effect on highly organized organisms is due mainly to the second cause. In both cases we find that the reacting systems are very sensitive to very weak natural EmFs. The actual functioning of these systems presumably involves electromagnetic processes which are modified in some way or other by external EmFs. The electromagnetic processes implicated in the regulation of vital processes in the organism will be the subject of the next chapter.

Chapter 12

ELECTROMAGNETIC FIELDS WITHIN THE ORGANISM AND THEIR ROLE IN THE REGULATION OF VITAL PROCESSES

As we have already seen, a high sensitivity to EmFs of various frequency ranges is exhibited by all species of living creatures and at all levels of functioning of the organism. On the other hand, physiological experiments have shown that the activity of any central or peripheral system of the organism and the effectuation of informational interactions in the organism depend on electromagnetic oscillations in a frequency range extending from infrared to low frequencies. Every year scientists are discovering more and more high-frequency components of such oscillations and new systems of "electromagnetic regulation" at the most diverse levels of organization in the organism — from the molecular level to the organism as a whole.

This has led to the hypothesis (Presman, 1964c) that EmFs of the most diverse frequency ranges play a significant role in the various processes of regulation and intercommunication with the organism.

12.1. Electromagnetic Regulation Systems in Living Organisms

In a wide variety of organisms we find regulation systems which depend on electromagnetic oscillations for their functioning. In this connection we can cite a few lines from Wooldridge's book (1963). "The nervous systems of the lower animals supply many examples of configurations of neurons that produce periodic output signals for the control of rhythmic bodily functions. . . . One interesting example of a neuronal oscillatory circuit is found in the lobster. It consists of a ring of nine interconnected neurons for the generation of the periodic electric impulses that control the heartbeat. . . . The song of a cicada is originated by an oscillator in the insect's brain. Here an interesting new circuit refinement

has been added — a subharmonic generator. The centrally origi-
nating frequency is 200 cycles per second, while the neuronal arrangement in the sound muscles drives them at only 100 times per
second. In many lower animals neuronal oscillator are responsible
for the rhythms involved in walking, swimming, or flying. The
swimmerets of a crayfish, wings of a locust, and crawling muscles
of an earthworm are centrally directed."

We could give more such examples by pointing to the more
complex system of electromagnetic regulation of the heart rate in
vertebrates, the diverse electromagnetic oscillatory systems of the
brain, which control both the rhythmicity of behavior and the rhythms of physiological processes, and so on. There is no need to
give an account of all these well-known control systems, but we
must mention one fundamental fact — the presence of two types of
such systems in organisms. This question has been fairly fully
discussed in Aladzhalova's monograph (1962) and we think it worthwhile to cite some of her generalizations and conclusions.

1. "Even at the level of the unicellular organism we can distinguish two control systems — a fast one and a slow one. The activity of the rapid system, in infusoria for instance, is manifested
in the beating of the cilia in response to random environmental factors and to short-term influences. The activity of the slow system
is manifested in a self-oscillatory process on the cell wall and the
excitability of the cells and the coordination of the beating of the
cilia vary in time with the rhythmic fluctuations of potential. The
purpose of the slow system is to ensure the stability of the entire
organism and to preserve its state under the action of a wide class
of external factors."

2. "From the rates of regulation in the central nervous system of warm-blooded animals we can distinguish fast and slow control systems. The first involves fast reactions to stimulation,
many of which, such as the orientational type of reaction, have already been investigated. The second, slow system assesses the
more or less systematically acting environmental factors and al-
ters the level of activity of the organism in connection with the regulation of homeostasis. The slow system acts on the parameters
of the fast system."

3. "One of the characteristics of the slow control system is
the fact that it does not react to an insignificant single (chance) ex-

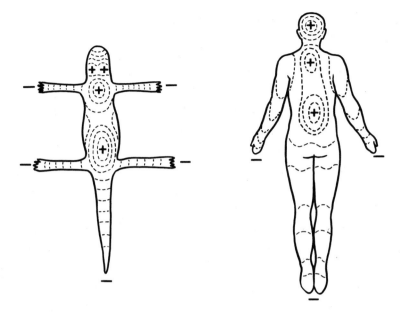

Fig. 93. Distribution of surface electric potential over body of lizard and man.

ternal disturbance. Its reaction to an environmental factor acting
more or less systematically takes several hours and can be directed
not only towards overcoming the produced changes in the internal
environment, but also towards active alteration of the level of activity with due regard to the possible effect of the new factor."

It is easy to see that our hypotheses on the difference in the
reactions of the central and peripheral systems to EmFs are in effect an application of these general conclusions of Aladzhalova to
a particular type of external factor — natural EmFs. In fact, the
fast reaction of peripheral systems to short-term exposure to EmFs
is a feature of fast systems, and the slow reaction of the central
systems and its two-phase dependence on the intensity and duration
of action of the EmF are features of the slow system. In addition,
we have discussed one other important property of these systems —
passive and active protection from EmFs which are alien to the
central systems.

Aladzhalova gives a list of various structures in which self-
oscillations accompanied by ultraslow fluctuations of potential take

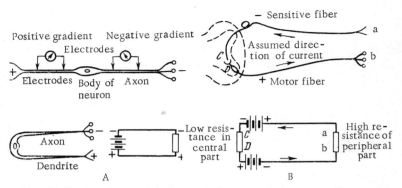

Fig. 94. Distribution of potential along spinal neurons and corresponding equiva
lent electric circuits. A) For neuron; B) for nervous center of spinal column.

place. The list includes paramecia and legume roots, the smooth
muscles of invertebrates and the skeletal muscles of the frog, the
dendrites and ganglion cells of mollusks, the nuclei of the hypothalamus, and the cerebral cortex.

This list, of course, does not exhaust all the electromagnetic
control systems found in living organisms. This is shown by the
discoveries made in recent years.

A slow electromagnetic control system was recently discovered by American scientists (Becker et al., 1962; Friedman et al.,
1962). An investigation of surface potentials in vertebrates showed
a distribution related to the position of the central nervous system
(Fig. 93), instead of the previously assumed dipolar symmetry of
the equipotential lines. It has been suggested that the spinal cord
generates a direct current which flows through the emergent nerves.
In fact, transection of the spinal cord led to a pronounced change in
potential on the limbs, while transection of the nerves going to the
limbs reduced the potential to zero. Further investigations by
means of the Hall effect showed that a direct current, due presumably to the movement of electrons along the nerve (whereas biocurrents in nerves are due to the radial motion of ions), passes along
the nerve fiber in the dendrite — neuron body — axon direction. It
was found that the sensory nerve fibers have a positive potential
at the peripheral end and the motor fibers have a negative potential.

Figure 94 shows these distributions and the corresponding equivalent electric circuits.

It was also found that injury to the limb led to the propagation of slow variations of potential along the spinal cord and a consequent active response of the brain approximately 2 sec after infliction of the injury. Finally, it was shown that the changes in the fronto-occipital potential in man during normal sleep, hypnoanalgesia, and total anesthesia were of the same type. Figure 95 shows the corresponding curves.

From these results the authors derived the following hypotheses:

1. In addition to the fast system of information transfer along the nerves the vertebrate organism has a slow electric control system involving the passage of a slowly varying current along the nerves.

2. This slow system regulates the activity of the fast system and, in particular, the rate of propagation of biopotentials. In addition, this system transmits slow information about pain and is probably implicated in psychic functions. Such a system was presumably formed at the early stages of evolution of the nervous system.

3. The slow system may control the general behavior of animals and through it the earth's magnetic and electric fields and changes in concentration of aeroions may act on the organism.

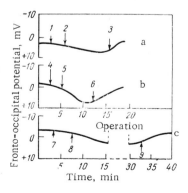

Fig. 95. Changes in fronto-occipital potential in man. a) During normal sleep; b) in hypnoanalgesia; c) under anesthetic. 1) Eyes closed; 2) deep rhythmic breathing; 3) arousal; 4) beginning and end of "hypnotic count"; 5) onset of hypnoanalgesia; 6) start of count for arousal; 7) injection of pentobarbital; 8) administration of nitrous oxide; 9) stoppage of administration of nitrous oxide.

Another electromagnetic control system has been discovered in the lobster ganglion (Watanabe and Bullock, 1960): A slow change in membrane potential in one of the large ganglion cells modulates the activity of the neurons (by altering the frequency of the discharges) located at a distance of several millimeters from the ganglion. This is a case of a long-range electromagnetic interconnection.

From the results of a harmonic analysis of the alpha rhythm of brain biopotentials Wiener (1963) expressed the following hypothesis: "We think that the brain contains several generators with frequencies close to 10 Hz and that within certain limits these frequencies may attract one another. In such circumstances the frequencies will probably collect into one or several groups, at least in certain regions of the spectrum."

An illustration of such "attraction" of frequencies is provided by the results of recent experiments with isolated heart-muscle cells (Harary, 1962). He found that while each cell exhibited its own individual rhythm of pulsation the assembly of cells pulsed with the same frequency, set by a "leading cell," which had the highest pulsation rate.

In addition to the discovery of new electromagnetic control systems, the theories relating to well-investigated systems of such kind are being developed. For instance, it has been found (Franke et al., 1962) that the human heart generates electrical oscillations in a much wider frequency range than was previously known, viz., from 30 to 700 Hz. In the case of pathological changes (pathological ischemia) the most pronounced changes are observed in the high-frequency part of the vibration spectrum. High-frequency components — in the range 200-500 Hz*— have also been discovered in the electrical activity of the brain (Trabka, 1963). These components are appreciably altered by barbiturate anesthesia.

There is an increasing amount of experimental and theoretical information on vibrational phenomena at the molecular and supermolecular levels in chemical and biological systems. These

------

* During the preparation of the book there appeared B. M. Nudel's paper (Third All-Union Conference on Neurocybernetics, September 7-12, 1965, p. 105) reporting the discovery of even higher frequency vibrations —up to 100 kHz.

questions were the topic of a recent special symposium (Vibrational Processes in Biological and Chemical Systems, 1967). Several papers (Frank-Kamenetskii, Shnol', Zhabotinskii, Sel'kov et al.) discussed the experimental information relating to vibrations in various chemical reactions and biological systems. Theoretical models of these vibrational processes and hypotheses relating to their mechnisms were discussed. The authors stressed that the investigation of such vibrational processes can play a very important role in the elucidation of the physicochemical nature of several biological phenomena — muscle contraction, regulation of cell division, the mechanism of the biological clock, photosynthesis, glycolytic metabolism, various enzyme-substrate reactions, and so on.

It is important to note that vibrations in chemical reactions and biological systems are always associated with electromagnetic vibrational processes (and in several cases are caused by these processes). There is no doubt that progress in the investigation of vibrational phenomena at the molecular level will greatly enlarge our view of the diversity of electromagnetic regulation systems and interconnections in living organisms.

Thus, experimental evidence of the sensitivity of organs, cells, and macromolecules to EmFs of different frequency ranges, of the generation of such fields in these systems and, finally, of some electromagnetic connections between them indicate the correctness of the hypothesis of the existence of diverse interconnections mediated by EmFs in living organisms. This is not a case of the already known methods of transmitting information along nerves by bioelectric impulses, but of special "radio communication"* between different elements and systems within the organism.

This hypothesis can provide a basis for an approach to an elucidation of the still unexplained physical nature of some interconnections between cells and macromolecules in the organism.

---

* The term "radio communication" in this case is arbitrary, since the interconnections apparently involve not only the radio-frequency range, but also EmFs of low and even infralow frequencies.

12.2. Possible EmF-Mediated Interconnections in Living Organisms

One of the important unresolved problems of biology is the explanation of the directional motion of cells in the body, and selective interaction between like cells at a distance. This problem has been examined in detail by Weiss (1959a, 1959b), who discusses the following questions:

1. ". . .why a free cell, which can extend in many directions, often advances steadily in one direction to the exclusion of others?" Discussing the hypotheses on the role of possible "tropisms" and gradients in these effects, Weiss points out that "it has never been possible to demonstrate just how a cell could translate such directional cues into actual convection towards or away from the source."

2. Referring to the fact that the cells "recognize" each other and their surroundings and can find their "proper destinations in the body even if they are deprived of their customary routes for getting there and that cells are grouped in types and actively preserve this grouping not only in the body, but even in experiments in vitro, Weiss poses the question: "How do mixed cell populations achieve this orderly reassortment? Do like cells "attract" each other? Or do they "recognize" each other only after chance encounters?"

3. Observing that there is still much that is obscure in the selective interaction between cells, Weiss concludes: "The only thing that seems clear is that these phenomena point to the same general area of specificity that covers immune reactions, enzyme-substrate interactions, pairing of chromosomes, fertilization, parasitic infection, and phagocytosis . . . . So, from phagocytosis and the uptake of macromolecules, through cell-to-cell contacts with specific "recognitions", up to the problem of "selective" adhesion of a cell to its substratum, one deals with a continuous spectrum of problems of the same nature."

4. Proposing a scheme of complex interactions in the entire organism (shown in Fig. 96) Weiss points out: "No outer agent can influence any of the inner shells except through the mediation of the

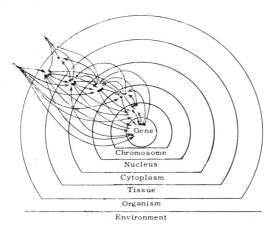

Fig. 96. Diagram of complex network of possible
interactions in organism.

shells in between, which may or may not modify that factor during
its inward passage. Conversely, products of inner systems may
not reach outer shells as such, but may be significantly screened
and altered in transit."

   If the described effects are considered from the standpoint
of the concept of diverse EmF interactions within the organism
and its interactions with environmental EmFs, then we can sketch
a picture which is very convincing in its simplicity and consistency.
In fact, we can picture the organism provided with diverse interconnections of such kind (in addition, of course, to the known diverse
neurohumoral connections), differentiated as regards their specific
"working" frequencies, intensity ranges, and method of coding.
Such interconnections may underlie not only the interactions between cells, but also specific interactions between macromolecules:
enzyme and substrate, antigen and antibody, DNA and RNA. Similar interconnections may be responsible for the control of protein
synthesis. In a recently proposed hypothesis regarding such control (New Biological Effects of R-F, 1959) DNA molecules are regarded as generators of radio-frequency signals, RNA molecules
as amplifiers, and enzymes and amino acids as effectors of signals
coded in various regions of the spectrum; the cell wall is believed
to act as a noise filter.

There are three fundamentally possible types of electromagnetic interconnections in the living organism: first, the known
scheme of interaction of the central control system with peripheral
effector and receptor elements, as occurs, for instance, in the nervous regulation of body functions; second, the automomous interconnections between elements, such as cells and macromolecules
(which are apparently also under central control) and; third,
signals from the control system simultaneously to different elements in the organism. The possibility of the existence of the last
type of connection was considered by Wiener (1958), who compares
it with a fire siren and calls it a "to-whom-it-may-concern" message. He suggests that messages of this kind can be transmitted
by the nervous system and humorally.

We think that this "emergency signal" is transmitted from
the central nervous system simultaneously to all the effector organs
by an "all-to-all" type of radio transmission and that it is associated mainly with emotional states of the organism.

As Simonov (1966) points out, in emotional states due to some
external source the organism is mobilized for "long-range action,"
directed either towards mastery of a useful object or to the removal of factors preventing satisfaction of a need, or to the avoidance
of danger.

It is known that in the case of "emotional mobilization" of
the organism the speed and strength of action of its organs (involved
in the performance of particular action) are much greater than the
normal "working" level. For instance, the coordinated action of
the muscles is much more rapid than usual and their contractile
power is enhanced. If we attempt to attribute this stimulation of
the organs as a reaction to signals brought to it through the nerves
from the central nervous system, we come up against a contradiction arising from the fact that the velocity of propagation of excitation through the vegetative nerves is only one-hundedth of the velocity of propagation through the motor nerves innervating the muscles (Knorre and Lev, 1963). It is the vegetative nerves, however,
which are responsible for activation of the processes leading to
the enhancement of the contractile ability of the muscles — release
of adrenalin by the adrenals, dilation of the muscular vessels, and
increase in the heart rate. In addition, it is known that the

heart rate is increased within a few seconds of stimulation of the vegetative (sympathetic) nerves.

Thus, we find a paradoxical situation: The muscles first receive the signal for enhanced coordinated action and the signals arrive much later at the organs responsible for the required enhancement of muscular activity. This naturally gives rise to the hypothesis of the existence of a system which sends "emergency" messages simultaneously to all the organs and which is not connected with the nerve network. This mode of signal transmission could ensure practically instantaneous total "mobilization" of the organism in emotional states. It seems probable that in these cases the central system transmits electromagnetic "to-whom-it-may-concern" signals at "emergency" frequency through the whole organism (in the same way as SOS messages are transmitted).

Thus, the hypothesis of the existence of diverse interconnections within the living organism by means of EmFs has at least an indirect experimental basis and can provide a starting point for an elucidation of the physical nature of some of the selective interactions observed between systems and elements of the organism. As regards the possible mechanisms of such interconnections, we can put forward only some general ideas.

12.3. Mechanisms of EmF-Mediated Interconnections within Organisms

We begin the discussion of such mechanism from the "bottom floor" — from the interconnections between macromolecules in the organism. Discussing the electromagnetic connection between macromolecules, Szent-Gyorgyi (1960) states that "... two molecules, the electrons of which are capable of a similar excitation, can thus act as coupled oscillators. In this case it is not necessary to have a material connection between the two, for the electromagnetic field couples them, provided their mutual distance is not too great (small compared with corresponding wavelength)." In addition, as mentioned above, there are theoretical and experimental grounds for inferring an electromagnetic interaction between macromolecules which might be regarded as informational, such as dipole—dipole interactions between identical and similar macromolecules due to fluctuations of the distribution of the proton and electron charge in the molecules (Vogelhut, 1960; Prausnitz et al., 1961). We have already referred to the various types of resonance absorption of EmFs by protein molecules and their aggregates, and such ability usually involves the possibility of appropriate electromagnetic radiation. Finally, resonance absorption of EmFs by biological molecules has also been demonstrated experimentally. All this suggests the existence of interconnections between macromolecules by EmFs of different frequencies, most probably in the ultrahigh-frequency and superhigh-frequency ranges (and, possibly, in the low-frequency range too).

Of interest in this regard are the recently expressed views of Neiman (1964, 1965a, 1965b) on the possible informational interaction at the molecular and atomic levels in such systems as DNA and RNA. In these systems, according to the author, there is an increased probability of determinative quantum processes due, first, to the increased probability of trigger action by a single interaction of atom and quantum and, second, to multiple interaction without increased probability of triggering by a single act. Such mechanisms are made possible by the fact that when there is no induced radiation the atom remains unexcited and the electromagnetic quantum also remains ready for action.

Neiman suggests that the total probability of induced radiation can be increased in three ways: first, by multiple action of the same electromagnetic quantum on the atom; second, by the action of a single quantum on a certain number of indentical and identically-excited atom, and third, by the action of several quanta on the atom.

In the estimation of the possible frequency range of electromagnetic interactions between cells two circumstances must be considered: On one hand, experimental evidence indicates that cells are sensitive to EmFs of various frequencies and, on the other, the penetration of EmFs into cells must depend on the frequency, since the impedance of the membrane and the absorption, due to polar molecules of water, depend on the frequency. An obvious compromise would be the range from tens of MHz, approximately, to 1000 MHz, in which the effect of cell membranes becomes insignificant, and the polar properties of water molecules are not effective. In other words, in this range the interconnection between the internal elements of different cells can be frequency-independent. A broad frequency band and a suitable coding method can ensure high reliability of intercellular interconnection, which is presumably effected through subcellular structures and at the molecular level.

Signal transmission from the central nervous system by the "all-to-all" type can be pictured as follows: The peripheral components, such as ganglia, endocrine glands, etc., have receivers tuned to "emergency" frequency which retransmit the signal at appropriate frequencies to the effector elements. The technical analogy of this is the retransmission of an important centrally transmitted broadcast by all the radio stations of a country.

Weiss's proposed scheme (1961a) of interactions in the entire organism can be applied also to the alien actions of environmental EmFs on the organism. In this case the alien action is electromagnetic interference, which is to some extent capable of disturbing the electromagnetic interactions within the organism. The protective mechanisms either do not transmit this interference to the electromagnetic control and intercommunication system, or modify them so that no disturbances take place. The action of EmFs of ultrahigh and superhigh frequencies on the central systems of the organism is limited primarily by the fact that the higher the frequency and the greater the size of the animal, the less the relative depth to which the EmF penetrates. At the same time, EmFs acting on the tissues of the organism are modified in them: The wavelength is reduced (by a factor $\sqrt{\varepsilon}$), the modulation characteristics are distorted, the average intensity is reduced with absorption in the tissues, and so on.

Finally, as already mentioned, there are grounds for postulation of a system of active protection based on a progressively increasing latent period of reaction to EmFs from the peripheral protective systems to the central control systems; hence, each peripheral system of protection manages to "warn" the system responsible for the next stage of protection and, finally, the central system, of the noxious action, thus giving the latter the opportunity to reduce its sensitivity to the acting EmFs.

The situation is different in the case of low-freqency EmFs, which penetrate into the body, and, to an even greater extent, in the case of infralow-frequency and constant fields, which can act on any elements of the organism. Here only active "warning" protection is possible.

On the other hand, the discussed scheme of multistage shielding must allow unhindered passage of regulating natural EmFs
(slowly periodically varying or high-frequency fields with particular
intensity modulation) into the "deep interior" of the organisms —
to the slow electromagnetic control systems.

Thus, the question of the nature of the regulating or disturbing effect of external EmFs on living organisms can be approached
from the standpoint of the existence in organisms of electromagnetic control systems and diverse EmF interconnections. From this
viewpoint the regulating effect of external EmFs is due to their direct action on electromagnetic control systems formed in the
course of evolution, and the disturbances caused by alien EmFs are
due to interference in the functioning of these systems or particular internal EmF interconnections.

Chapter 13

ROLE OF ELECTROMAGNETIC FIELDS
IN INFORMATIONAL INTERCONNEC-
TIONS BETWEEN ORGANISMS

Various kinds of long-range informational interconnections
between animals have been known to biologists for a long time. Animals could not exist if they were unable to exchange signals by
which the female calls the male, the mother her children, and individuals warn each other of danger, report the location of food, and
so on. We know the physical nature of many forms of this signal
transmission — sonic (and ultrasonic), photic, and by means of
smells (see, for instance, McElroy and Seliger, 1962; Rait, 1964;
Chauvin, 1965).

In addition, new physical agents capable of carrying information have been discovered in recent years. For instance, Brown
(1963c) discovered that planarians can orient themselves relative
to weak gamma radiation (only six times more intense than the natural level) and can distinguish its source, while some experiments

with ants indicate the possible existence of intercommunication based on ionizing radiation (Khalifman, 1965; Marikovskii, 1965).*

However, there are various manifestations of informational interconnections between animals which are still of a puzzling nature: What signalling system underlies the simultaneity of the movements of birds in a flock and fish in a shoal? Why do paramecia, which continuously move about rapidly in all directions, never collide with one another and when they come too close stop instantaneously and alter their direction sharply? How do animals transmit signals warning of danger when they are at very great distances from one another? How does a dog find the shortest way to its master from a great distance when there is no possibility of orientation by means of the known sense organs? We could ask more questions of this kind and also point out the unconvincing nature of the explanation of some interactions as, for instance, the attraction of the male by the female by smell over distances of several kilometers.

Signals transmitted in the animal world by methods of as yet unknown physical nature have recently been called b i o i n f o r m a - t i o n . There are grounds for believing that in several cases such signal transmission is effected by EmFs of various frequency ranges. This is borne out, firstly, by the high sensitivity of animals of the most diverse species to EmFs, and especially by the fact that EmFs can act as a conditioned stimuli for the elaboration of conditioned reflexes. Secondly, it has been found that people exposed to EmFs experience various sensations and that some animals have special EmF receptors. Thirdly, EmFs of various frequencies have been recorded in the vicinity of isolated organs and cells, as well as close to entire organisms.

13.1 Sensations and Unconditioned
Reflexes Due to EmFs

As long ago as the end of last century D'Arsonval (1893) discovered the phenomenon of phosphene (sensation of light flashes

---

* E d i t o r ' s   n o t e : Other evidence suggested that responsiveness of planarians and mice to gamma fields may be continuously exhibited even at the natural ambient background levels (Brown and Park, 1965; Brown, Park, and Zeno, 1966).

in the eye) due to the action of an EmF. Further investigations (Danilewski, 1905; Thomson, 1910; Barlow et al., 1947; Mogendovich and Skachedub, 1957; Valentinuzzi, 1962; Solov'ev 1963; and others) showed that this effect could be produced by a constant magnetic field or by the magnetic component of an EmF and it was called "magnetophosphene." It has recently been reported (Jaski, 1960) that people exposed to an EmF in the frequency range 380–500 MHz are subject to visual hallucinations: The subjects invariably associated these sensations (in 14 out of 15 tests) with the same point in space.

Sound sensations due to the action of an EmF were first observed in 1956 and were subsequently investigated thoroughly by Frey (1961, 1962, 1963a), who managed to establish the following features:

1. People exposed to a pulse-modulated EmF hear various sounds (buzzing, knocking, or whistling), depending on the modulation. The sound source seems to be somewhere in the region of the occiput. Experiments involving screening of the head showed that "radiosound" could only be perceived in the temporal region.

2. "Radiosounds" are very similar to the sounds produced by amplified random oscillations (noise) of the envelopes of the modulation pulses with frequencies below 5 kHz cut off and maximum extension of the band towards higher frequencies.

3. Surrounding noise of up to 90 db does not dispel the "radiosound," although it reduces sensitivity to it. The use of ear plugs enhances the "radiosound" effect.

4. There are certain threshold intensities below which there is no effect. Their values depend on the EmF parameters, the main ones in this respect being the strengths of the electric and magnetic components in the pulse, and not the power flux density.

Frey suggests that the EmF is preceived directly in the auditory nerves and in the cells of the auditory zone of the cerebral cortex by various "detectors" and that other sensations, in addition to "radiosound", can be produced by EmFs, depending on the irradiation conditions.

Wieske (1963) later observed similar effects due to the action of low-frequency EmFs — from 60 Hz to 15 kHz (with maximum sensitivity at 3 kHz) — and also in response to the application and removal of an electrostatic field. Wieske suggests that the EmF induces weak currents in the inner ear and these excite the auditory cells or auditory nerves.

A tactile sensation in man due to a pulsed electric field has been reported (Noval, 1963). The subject, situated inside a solenoid, experienced jolts — on both sides of the spine in the femoral region when lying on his back, in the epigastric region when lying on his side and, in the stomach region when lying on his stomach. Frey (1963b) also described tactile sensations — itchiness or soreness on the skin of the face or forearm — in people situated close to the antenna of an ultralong-wave (14.7 kHz) radio station.

Lastly, we mention Turlygin's experiments (1937), in which exposure to centimeter waves of very low intensity caused a sensation of drowiness and weakness in human subjects.

EmFs can act as a conditioned stimulus, as we have already mentioned, in experiments with man (Plekhanov, 1965) and in experiments with animals (Malakhov et al., 1963, 1965a).

We encounter a new kind sensitivity — specific reception of an EmF — in fish. This kind of sensitivity was first discovered only in fish with electric organs, which can react to electric field pulses of very low strength — 10^{-6} V/cm (Lissman, 1958; Lissman and Machin, 1958; Fessard and Szabo, 1961; Szabo, 1962), but it has recently been found that other species of fish are sensitive to an electric field. This sensitivity is different in different kinds of fish. For instance, Bodrova and Krayukhin (1965) found that the ray was only half as sensitive as the sea perch; the pike-perch is half, and the eel pout one-third, as sensitive as the pike; in a test of similar-sized fry of 13 species of sea fish, mackerel were found to be the most sensitive. The relationship between the electrosensitivity of fish without electric organs and the pulse length and repetition rate was recently investigated (Fig. 97, Lissman and Machin, 1963).

If we compare the data on the generation of pulsed fields by fish and their sensitivity to such fields we have adequate basis to postulate the existence of an informational interconnection by means of EmFs.

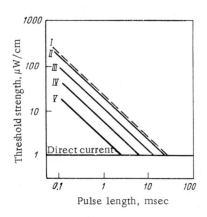

Fig. 97. Threshold strengths of electric pulses causing a conditioned-reflex reaction in fish as a function of pulse repetition rate. I) Single pulses and 10-20 pulses/sec; II) 50 pulses/sec; III) 100 pulses/sec; IV) 200 pulses/sec; V) 500 pulses/sec.

Electrosensitivity in fish seems to be associated with certain sensations, but any attempts to describe them have been absolutely fruitless (as, in general, the anthropomorphic approach to sensations in animals). Here we are dealing with the reception of a physical agent, a process which has no analog in man, as is the case with other kinds of reception in animals, such as the perception of ultraviolet rays by the bee, infrared rays by the snake, and so on. We introduce these trivial considerations merely to underline the following important point: All the reactions of animals to EmFs which do not involve sense perception can be compared to some extent with corresponding reactions in man (for instance, a change in physiological processes), but there are no parallels in cases where the reception of the EmF involves some kind of sense perception.

Thus, we can infer that the ability of animals to receive EmFs and the induction of particular sensations in man by EmFs are due either to the activity of undiscovered sense organs or to still unknown properties of the usual sense organs or combinations of them.

13.2 EmFs Created Close to Cells

and Organs

As far back as 1947 Lorente de No found that an electric field was formed around an excited nerve situated in a conducting medium. Two years later Burr and Mauro (1949) detected a field extending to a distance of several millimeters in air space around a nerve.

In 1948 Krayukhin, and subsequently Seipell and Morrow (1960), managed to detect with an induction transducer a voltage due, in their opinion, to the magnetic component of an electromagnetic field close to an excited nerve. Further investigations of this effect gave contradictory results: Some investigators (Gengerelli, 1961; Gengerelli et al., 1964) confirmed this effect, while others (Khvedelidze et al., 1965) could not find it. A possible reason for these contradictions is that the recorded quantities are very small and their measurement is a very difficult methodological and technical problem. However, extremely weak magnetic fields (of the order of millionths of an oersted) arising in the vicinity of the human heart have recently been recorded (Baule and McFee, 1963). The method used in these investigations was so reliable that it has led to the development of a practical apparatus (using induction transducers) for "magnetocardiography."

Investigations in which high-frequency radiation from human muscles has been observed (Volkers and Candib, 1960) are of great interest. Low-frequency and high-frequency radiations in the range from tens to 150 kHz were recorded during muscular contraction (by direct contact). The greatest effect was observed in small muscles and a moderate effect in the gastrocnemius. The head muscles did not emit any radiation. Various methods of measurement (by means of amplifiers with a very low noise level) — broad-band, narrow-band, and at specific frequencies — revealed variations in the nature of the radiations depending on the age and state of health of the subjects. The authors suggest that there is a higher-frequency radiation of muscles.

Similar investigations were conducted later by a long-range method with an antenna placed at a distance of 1 cm from the object (Malakhov et al., 1965b). The amplifiers were designed for two frequencies (3 and 150 kHz) and had passbands of 1 and 10 kHz, respectively, and sensitivities of 0.2 and 0.1 μV. The experiments were conducted on frog gastrocnemius and human forearm muscle and the differences in the signals in the contracted and relaxed state of the muscles were estimated. Contracted forearm muscle showed weak emission at 3 kHz (0.1 μV in the antenna) in the form of random pulses about 1 μsec long; relaxed muscles did not emit. No radiation at 150 kHz could be detected. The authors suggested that the spectral density of the muscle biopotentials decreases with increase in frequency and they investigated the spectra of biopo-

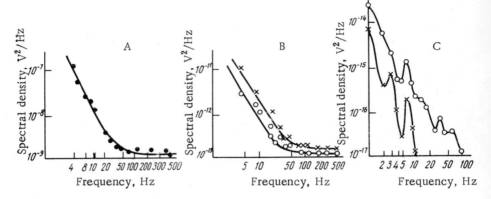

Fig. 98. Spectral characteristics of biopotentials of excited forearm muscles (A) and visual lobes of dragon-fly brain (B), and frog brain (C)   Circles, dark-adapted eye; crosses, eye illuminated.

tentials on various objects by means of a spectrum analyzer (frequency band 1 kHz, sensitivity not less than 10^{-16} V²/Hz). Figure 98 shows the obtained spectral characteristics.

The most interesting and convincing results were recently obtained in Leningrad State University (Gulyaev, 1967; Gulyaev et al., 1967). A specially designed probing amplifier made it possible to detect the EmF formed around active nerves, muscles, and heart of the frog, and the EmF produced by human musculature and heart. The authors called the recorded fields the "electroauragram." The electroauragram of isolated frog nerve was detected at a distance of 25 cm from it (where the voltage was 1 mV), that of isolated frog muscle and heart at a distance of 14 cm, and that of human heart and muscle at 10 cm. EmFs created during the flight of the bumble-bee and mosquito have also been detected.

The generation of electric pulses by fish with electric organs has been investigated for a long time. It has now been established that such fish emit pulses of low voltage (of the order of 1 V), either in the form of separate low-frequency bursts, or continuously. The frequency, shape, and length of the pulses are characteristic of each species of fish (Bullock, 1959). The usual frequency range is 60–400/sec, but in some cases exceeds 1000/sec. The pulse duration ranges from 10 msec to less than 0.2 msec. The mechanism of action of the electric organs, which consist of sets of electric plates, has been fairly well investigated (see Bennett's review, 1964).

The functional significance of these organs, however, is still obscure.

The very fact of detection of an EmF around isolated cells and organs and in the vicinity of entire organisms indicates the possibility of exchange of information between animals by means of EmFs. These first results must be regarded as very promising, since the conditions in which they were obtained were not, in general, optimum.

First, the fields were recorded in the absence of sufficiently definite information about the parameters of the investigated EmFs and on the basis of a rough estimate of the actual frequency range, the nature of the coding (modulation), and the intensity of EmFs produced by biological systems; second, in experiments with entire organisms the result obtained was due to the addition of non-coherent EmFs of elementary generators (nerve and muscle cells), and the total intensity was proportional to the number of generators, whereas information transmission is presumably effected by coherent vibrations, where the total intensity is proportional to the square of the number of generators; third, it seems likely that information can be transmitted in a wide band of frequencies, when the elementary generators simultaneously transmit the message at different frequencies.

Below we discuss the question of what kinds of bioinformation involve coherent transmission and what require broad-band transmission. We also discuss possible schemes of EmF-mediated interconnections in the animal world.

13.3 Bioinformation Transfer

and EmFs

The cited experimental information indicating the ability of the animal organism to generate and perceive EmFs of various frequency ranges can be regarded as indirect evidence of the existence in the animal world of EmF-mediated informational interconnections. In addition, as we have already mentioned, there are many examples of information interconnections of as yet unknown physical nature between animals and it seems probable that signal transmission is effected in these cases by EmFs.

We will now discuss the following questions: 1) the possible types of bioinformation transfer by means of EmFs; 2) the basic approach to investigations aimed at direct detection of such bioinformation transfer, and 3) model systems which might help to determine the corresponding mechanisms of transmission and reception of information.

It is obvious that these questions can be answered only by proceeding first from an investigation of the b i o l o g i c a l r e - l a t i o n s h i p s underlying bioinformation trasfer. This means that it is necessary 1) to undertake an all-round investigation of the behavior of animals in natural conditions or in conditions approximating as closely as possible to natural; 2) to consider the variation of the physiological and emotional state of animals with their age, seasonal conditions, reaction to surroundings, etc., and 3) to make the correct choice of species of animal for the investigation of a particular kind of bioinformation transfer.

This approach to the investigation of bioinformation transfer will undoubtedly entail many difficulties. However, it would be impossible otherwise to determine the optimum biological conditions in which such transfer is manifested and without this it would be impossible to conduct effective physical investigations for its direct detection and for the investigation of the parameters of the acting EmFs and the mechanisms of transmission and reception of information.

A consideration of the special features of animal behavior which might be attributed to interconnections effected by EmFs reveals four types of bioinformation transfer. We will discuss in turn their main features, possible models, and methods of detection.

The f i r s t t y p e is the bioinformation transfer responsible for the rapid coordination of the activity of individuals in groups and communities of animals. Such an informational interconnection presumably underlies such phenomena as the simultaneous changes in the direction of motion of birds in flocks and fish in shoals, the rapid coordination of the activity of social insects at relatively short distances from one another, the absence of collisions among randomly moving infusorians, and so on. In all these cases the connection is made over relatively short distances and can be effected by single weak EmF signals carrying a small amount of information. We will discuss possible model schemes of such connections.

Zoologists believe that birds in flocks and fish in shoals perform simultaneous maneuvers in response to certain signals from a leader. If this signal transmission is effected by EmFs, it can be done in two ways: Either the leader is a sufficiently powerful generator of EmFs for the transmission of signals "to all in the flock," or relatively weak signals from the leader are retransmitted from individual to individual. Such a connection can be effected in two ways: Either the signals arrive at the brain centers which coordinate the flying and swimming movements and these centers send an appropriate "command" along the nerves to the organs responsible for effecting the maneuver, or the signals go directly to these organs and cause their reflex action. It is well known that a particular motor act by an animal can be produced either by electric stimulation of the appropriate region of the brain, or by direct stimulation of the neuromuscular apparatus.

Infusoria swimming about rapidly in all directions never collide with one another, but make sudden stops and alter their direction sharply in time. Our experiments with paramecia, however, showed that such maneuvers can be produced by exposure of infusoria to EmF pulses. The paramecia responded to each pulse by a sudden stoppage of movement and simultaneous orientation of the body parallel to the electric lines of force. It seems probable that such motor actions in natural conditions can be brought about by electromagnetic signals produced by the periodic changes in the electric potential in the membrane of the paramecia due to beating of the cilia (see Chapter 9).

In all the considered cases the signals are probably transmitted simultaneously by all the nerve-muscle elements implicated in the motor act of the animal. As was mentioned above, any forms of motion (swimming, flight, crawling, etc.) give rise in these elements to coherent electromagnetic vibrations synchronized by a control center (in infusorians they are synchronized in the "neuromotor" apparatus itself). The signals are obviously received by spatial summation either in the elements of the center responsible for control of the movement or in the elements which perform the motor act. By this method information can be transmitted by relatively weak EmF signals and can be sufficiently reliable against a background of interference and noise.

The following experiments might be carried out for the direct detection of bioinformation transfer of the first type:

1. Observation of the effect of EmFs of various frequencies on the maneuvers of birds in flocks (for example, in flights close to radio stations or in the zone of radiation from experimental transmitters), on the motion of fish in response to exposure to constant or pulsed electric fields, on the behavior of infusoria in an EmF and, finally, the reaction of social insects to EmFs (recall the effect of SHF fields on the behavior of ants, as described in Section 7.1), and so on.

2. An investigation of the behavior of social insects with various methods of screening of individuals or groups of them [as was done in Leconte's experiments, described by Chauvin (1965) and in Marikovskii's experiments with ants], thus creating conditions in which the insects can "select" between natural conditions and zones with an artificial EmF [for instance, between the plates of a capacior, as in Edwards' experiments (1960a)], and so on.

3. If these methods provide even an approximate estimate of the frequency range of electromagnetic signal transmission, then attempts can be made to detect it directly by means of a sufficiently sensitive apparatus. It would be better to begin investigations of this kind with social insects or unicellular organisms, since in these cases it is relatively easy to provide the necessary shielding from external interference and to place the antennas at short distances from the object.

The second type of bioinformation transfer is the relatively slow signal transmission which might be regarded as the explanation of the hitherto baffling ability of many animals to find their way to one another over a long distance. This kind of ability is found, for instance, in insects (the female summoning the male from a long distance) and in birds (the "homing" instinct). In these cases information has to be transmitted over considerable distances and, hence, fairly strong signals and high sensitivity of perception of such signals are necessary.

It is characteristic that such navigational powers in animals depend on their emotional state. The female calls the male only in the nuptial period, when they are both imbued with the emotion of sexual attraction. The "homing" instinct is manifested mainly

at the time of feeding of the young and, hence, is also associated
with emotional excitation: with the sensation of hunger or alarm
in response to threatened danger in the case of the young birds and
with the urge to feed the young birds and protect them from danger
in the case of the parents.

We have already mentioned (Section 12.2) that when the organism is in an emotional state electromagnetic signals can be transmitted from the center and can lead to simultaneous mobilization
of all the effector organs for action in relation to the source of
emotion. In this case the signal from the center may give rise to
coherent electromagnetic vibrations of the same frequency or vibrations of different frequency reproducing this signal in the effector
elements. If this is true, intense transmission of information either
by an EmF of particular frequency or at different frequencies simultaneously becomes possible in the emotional state. We can assume too that in this state the animal organism becomes more sensitive to the reception of the two kinds of messages. As already
mentioned, the high sensitivity to EmF signals may involve the
mechanism of spatial summation. Since the transfer of bioinformation of the considered type takes a long time, the mechanism of temporal summation of multiply-repeated signals seems probable.

If we try to find a technical analogy to the navigation of animals in search of one another we can say that they act as "living
radio beacons." One animal continuously sends out EmF signals,
by which the other orients itself and finds its way. Signal transmission of a similar kind presumably underlies other kinds of navigation in animals. In the case of "homing" it may be electromagnetic radiation from the young birds (or adult birds, nesting "neigh-
bors"), and in the case of migration of elvers (from the Sargasso
Sea to rivers) it may be radiation from adult individuals situated at
the destinations.

The following methods might be used for the experimental
detection of the second type of bioinformation transfer.

1. Experiments with animals screened from one another;
a comparison of the number of encounters of male insects with
screened and unscreened females; a comparison in the behavior of
birds (signs of "unease") when their young are threatened with danger (not artificially created danger, but natural danger, to which
the birds react emotionally), and so on. Care should be taken to

ensure that the experimental conditions approximate as far as possible to natural conditions and do not introduce artifial interference.

2. Comparative investigations of the interconnections between emotionally excited animals in natural conditions and subjected to EmFs of a particular frequency range. In the investigation of "homing," experiments of such kind can be carried out with the nest containing the young birds exposed to the action of the EmF accompanied by a threat of danger or in the absence of such a threat. It would be worth while to carry out such experiments with insects during their nuptial period.

The third type of bioinformation transfer is the slow exchange of information by EmFs between individuals of one population or one species. Such exchange of information might be regarded as one of the possible mechanisms of regulation of development of a population and intraspecific development (qualitative and quantitative).

The existence of intrapopulation and intraspecific regulation is indicated by Chauvin (1965), who regarded it as a "communal nervous system" — a set of interconnections between individuals to ensure a higher level of their joint activity. He recalls a hypothesis put forward in 1925 by Uvarov to the effect that the change in color of the green locust to black in association with the black locust is due to some unknown stimulus emanating from the latter. He goes on to state that there are grounds for inferring the existence of some mechanism which automatically controls the number of a species irrespective of the available amount of food and concludes that there is "some mean population density, different for each species, which inevitably brings into action a mysterious regulatory mechanism which, acting through the adrenals and hypophysis, first inhibits, and then completely suppresses, reproduction". Is this not a case of regulation by EmFs?

To these hypotheses we can add that of the English zoologist Hardy (1949). He postulated that between individuals of the same species there is some long range interconnection which plays an important role in the formation of species behavior and in the general direction of evolutionary development within a species. This hypothesis, which admittedly is purely speculative, is another attempt to bridge the gap between the preformational and epigenetic theories of heredity.

We will consider this hypothesis from the standpoint of an electromagnetic informational interconnection between animals. We have described the genetic effects of EmFs and have mentioned the theoretical possibility of an electromagnetic interconnection between nucleic acid molecules. This suggests that the exchange of genetic information within a population (which has a set of genetic information — the "gene pool") takes place not only by mating, but also by electromagnetic interconnections between individuals.

All these forms of bioformation transfer of the third type can be effected by repeated transmission of EmF signals over a long period. Information transmitted in this way gradually accumulates in the receiving organisms. It is rather difficult at present to conceive how the postulated existence of such information transfer might be verified. We can only suggest that the methods employed by zoologists to observe the development and behavior of individuals isolated from their fellows in the population, with the addition of electromagnetic screening, might be used.

The fourth type of bioinformation transfer might be regarded as one of the possible factors governing the behavior and development of animals in groups and communities in their interaction with environmental EmFs.

An illustration of this type of bioinformation transfer is the experimentally established fact that although individual birds have relatively low orientational ability, the power of the community in this respect is very high. Observing this fact, Naumov and Il'ichev (1965) state: "The community in this case frequently emerges as a mediator between the individual and various environmental factors, like a whole "organism" with interdependent interconnections." We think that this idea may prove to be very fruitful in attempts to explain the hitherto baffling mechanism of animal orientation and navigation.

We may well encounter "mediation" of this kind when we consider the causes which induce animals to migrate. There are contradictory opinions on this and we cite only Chauvin's views (1965): "No one has yet managed to explain why the locust chooses a particular direction, why it arrives, or why it leaves. The first hypothesis proposed was, of course, the simplest: The locust (and all other migrating animals) leaves a locality in search of food. But this is quite untrue for the locust and for all other migratory animals. On the contrary, the locust can leave a quite unused rich feeding ground and fly off to certain death in the desert or plunge in hundreds of billions into the depths of the sea." Further on he writes: "The fever which grips mammals during migration appears to me to be a manifestation of some profound disorder of the neuroendocrinal system, which has no definte or direct connection with food, but is possibly due to some unknown pronounced changes in meteorological conditions. Some authors have spoken of cycles of solar activity in this connection..."

The causes of animal migration might be considered in the light of the above-discussed information and hypotheses on the regulating effect of periodically varying natural EmFs and the disturbing influence of spontaneously increasing EmFs. It is well known, for instance, that many kinds of birds migrate in spring from tropical regions to central and northern latitudes. The tropical zone, however, is the main center of thunderstorms — the number of thunderstorms is particularly high in spring and summer. There is much less thunderstorm activity in central and northern latitudes. In the USSR, for instance, it varies from 40-50 days in the year in the Caucasus to five days per year on the northern coastline of the country.

This comparison suggests that thunderstorm activity may be one of the significant reasons which induces birds to migrate in spring from the tropics to the central and northern latitudes. We have already pointed out that EmFs have very pronounced adverse effect on the development of the organism from the embryonic state to the formation of the adult individual. This harmful effect of EmFs produced by atmospheric discharges may be the reason why birds migrate to rear their young in regions with little thunderstorm activity. The effects will be less in the case of mammals, which develop in the mother's womb, and in the case of animals which are naturally protected by earth or sea water, for instance, in the embryonic stage. Perhaps the baffling spawning migration of eels to the great depths of the Sargasso Sea is also to provide special protection for the embryos from the influence of atmospheric fields.

The concept of the informational functions of EmFs in living nature may also prove fruitful in the approach to the problem of the mechanisms controlling the size of animal populations, as well as to the problem of dynamic equilibrium in the biosphere.

It is obvious that the factors on which the size of a population depends are diverse and in several cases still unknown and that the size of a population is regulated by feedback mechanisms — the size of the population influences its rate of growth (Ehrlich and Holm, 1963). We will consider these questions in the light of the postulated existence of EmF interconnections creating some over-all "electromagnetic background" in the population.

We saw above (Section 8.2) that experiments with animals of various species revealed inhibition of development by EmFs. One can surmise that on the attainment of some critical population density the intensity of the electromagnetic background increases so much that it becomes a factor inhibiting multiplication. Developing this hypothesis, we can postulate that EmF interconnections also regulate the relationship between the sizes of populations of different species. Here each population emerges as a single "organism," interacting with other similar "organisms" and with environmental EmFs.

We will now return to the question of animal orientation and navigation. We will consider this question in the light of the above-described experimental information on the orientational effect of the earth's electric and magnetic fields on animals. The relatively poor orientational ability of separate individuals is greatly increased in a community of them. Here we apparently encounter a self-organizing, self-correcting system, which is very superior to its constituent elements or, in other words, the property of "superadditivity" of living systems, of which von Foerster (1962) wrote.

How can the hypothesis of the existence of the fourth kind of bioinformation transfer be experimentally verified? This is a much more difficult problem than in the previous cases, but we can outline some methods of investigation.

1. Observations of the behavior of animals in natural conditions, such as: comparison of the numbers of birds leaving particular wintering regions with the characteristics of thunderstorm activity in these regions; a search for a correlation between the size of populations and spontaneous increases in the intensity of natural EmFs. and a similar approach to sudden migrations of animals.

2. Observations of the development of young birds exposed to artificially created electromagnetic interference in their nesting sites and observations of "homing" in the presence of electromagnetic interference in natural conditions or in experimental cages.

3. A comparison of the effect of electromagnetic interference on the orientational ability of individual animals, groups, and communities; an investigation of the orientational ability in experiments involving screening.

13.4. Parapsychological Investigations

In discussing the question of exchange of information by means of EmFs in the animal world we cannot avoid the "problem of parapsychology," which for decades has been the subject of numerous investigations and papers in specialist and popular-scientific journals and which is still a subject of very keen discussion.

We will not set out to analyze here the arguments of the advocates and opponents of parapsychology, but will merely try to assess objectively the present state of this question and put forward some views on the problem itself and the methods of investigation used.

The investigation of parapsychological phenomena began with the founding of the Society for Psychical Research in London in 1882. The problem of parapsychology includes the whole range of "superpsychic" human powers — the ability to transmit and receive mental information without the aid of known sense organs ("telepathy"), to determine the nature and location of objects inaccessible to sensory perception ("telesthesia"), to move objects by mental effort ("telekinesis"), to divine the past ("retroscopy"), and to predict the future ("perescopy"). Various spiritualist "phenomena" are also often included in the field of parapsychology. If we confine ourselves to investigations which can be regarded as scientific to some extent then we need only consider telepathy and telesthesia.

The methods of parapsychological investigations of telepathy and telesthesia reduce mainly to the following (Vasil'ev, 1962a, 1962b).

1. Observations and analysis of cases of "spontaneous telepathy," where people suddenly experience a sensation of anxiety

about some close friend or relative without receiving any news at that particular moment. The feeling is either a vague sensation of anxiety or a more definite sensation — a conviction that someone close has fallen ill, or suffered some misfortune, or has died.

2. Experiments in which a "percipient" tries to carry out a mental instruction from an inductor when they are in contact with one another (the percipient holds the inductor by the hand) or not in contact.

3. Experiments with Zener cards (each of which has a picture of one of five geometrical figures) in which the inductor tries to convey mentally to the percipient a randomly chosen sequence of cards.

4. Experiments in which the percipient guesses the nature of objects hidden from view.

Investigations of such kind have been the subject of several papers and even monographs (Rhine and Pratt, 1957; Soal and Bateman, 1954; Ryzl, 1926; Mancharskii, 1964; Vasil'ev, 1962a, 1962b; and others). These publications describe numerous cases of telepathy and telesthesia, which (in experiments with "specially sensitive" subjects) were reputedly observed when the inductor and percipient were separated by very large distances (up to several thousand kilometers!) and when one of the subjects was installed in a screened room.

The results of these experiments led parapsychologists to assert that telepathic communication is unaffected by distance and material barriers. This gave rise to hypotheses of special "biological forms of energy," "psi-fields," "biological quanta," and so on (Wasserman, 1956). Such highly penetrating agents as neutrinos and gravitational fields have also been considered (Ruderfer, 1952). It has also been postulated that telepathic communication is of an electromagnetic nature and occurs at extremely high frequencies, from 10^{19} to 10^{29} Hz (Bibbero, 1951), or in a very wide band of frequencies, from infralow to superhigh (Mancharskii, 1964), or on superlong radio waves (Kogan, 1966).

The electromagnetic theory of telepathic communication appears to have been confirmed experimentally. The first experiments were carried out by the Italian researcher Cazzamali (1925, 1928, 1929, 1935, 1936, 1941). He put the subject into a screened chamber containing radio receivers for the frequency range 3 to 400 MHz and a broad-band generator (60-400 MHz). Cazzamali claimed that when the subject was in a state of strong emotional excitement, electromagnetic radiation in the whole investigated range could be detected by the "beating" method. Turlygin (1942) later reported radio emission from the human brain. He investigated the reaction of a subject (sweating) to nonverbal hypnotic suggestion with the hypnotist screened by reflecting screens and diffraction gratings. The author concluded that the hypnotist emitted radio waves in the range 1.8-2.1 mm.

Many physicists have a sceptical attitude towards Cazzamali's experiments and believe that the observed signals were the result of various side effects. For instance, Petrovskii (1926) suggested the possibility of self-excitation of the radio receivers due to a change in the position of the subject's body, which can be regarded as a passive antenna. Holman (1952) believed that the emission of the transmitter might excite electromagnetic vibrations in the chamber as in a cavity resonator, and the change in the electric parameters of the subject's body in the case of emotional excitation might be the reason for modulation of these vibrations.

There have been no reports of the reproduction of Cazzamali's experiments on electromagnetic emission of the brain. His other experiments (1941), however, the production of hallucinations in subjects due to EmFs of the investigated range, were recently successfully reproduced (Jaski, 1960).

Thus, a considerable amount of experimental material on telepathy has been gathered and there is a basis for discussing the physical nature of telepathic communication. But are the results of telepathic investigations sufficiently reliable? Let us see what researchers on telepathy say on this subject themselves.

At a special symposium of parapsychologists held in 1956 (Ciba Foundation Symposium on Extrasensory Perception) the following doubts were expressed.

1. The results of experiments with Zener cards might be attributed to the inadequate nature of the employed methods of statistical analysis of the data.

2. There have been several cases of "false telepathy," involving coded transmission of information to the subject from his

assistant or detection by the subject of involuntary external signs in those taking part in the experiment.

3. A fact which should engender precaution is that the results of experiments on telepathy cannot usually be reproduced.

Such doubts led the American research worker Hansel (1966) to analyze thoroughly the methods and results of practically all the known experiments on telepathy carried out in the last 80 years (up to 1966). Having discovered various methodological errors in early telepathic experiments he repeated these experiments and carefully explained where and in what conditions the error occurred. In several cases Hansel repeated recent experiments (together with their authors) in an attempt to reproduce the experimental conditions as accurately as possible. This work led Hansel to the categorical conclusion that absolutely all the results of telepathic investigations are unreliable and could be attributed to faulty experimental procedures or incorrect statistical treatment of the data and that in most cases there was conscious or unconscious deception. Hansel was highly critical of some "sensational" results and accused the authors, Rhine and Soal, for instance, of deliberate distortion of the experimental data.

Despite such categorical conclusions, Hansel did not reject the possibility of telepathic phenomena and pointed out that it was essential to obtain convincing evidence from correctly designed and reproducible experiments.

We are inclined to agree with Hansel's conclusion regarding the unreliability of the results of telepathic investigations (leaving aside the accusation that research workers deliberately distorted the experimental data) and also with his recommendation that "parapsychological" methods of investigation should be replaced by better methods. However, we think that not only the techniques are unsatisfactory, but also the choice of object — man. The general biological approach to the problem of bioinformation transfer in its evolutionary aspect leads to the same conclusion.

The ability of animals to exchange signals over a distance without the help of known sense organs can be regarded as an attribute which has been developed in the process of evolution to ensure a better chance of survival of the individual and its success in the struggle for existence and for preservation of the species.

As man, however, has invented more and more diverse and improved means of artificial communication, his ability to transmit such signals must have become weaker and weaker and, finally, have been lost altogether. In other words, people today will exhibit this ability very rarely — as an atavistic attribute. This is why man is the most unsuitable object for the investigation of bioinformation transfer. This subject will have to be investigated in experiments with animals in which special features of their behavior are investigated by the methods described above.

It is quite possible, however, that if experiments with animals are successful, better investigations can be undertaken to detect the atavistic power of people to transmit bioinformation and, perhaps, methods of enhancing this ability might be found (for instance, by appropriate training or the use of certain stimulators).

As we have pointed out, bioinformation transfer between animals situated at great distances from one another is usually associated with particular emotional incentives and states.

If people can transmit bioinformation then this ability should also be manifested in emotional states. Hence, the above-mentioned parapsychological methods of investigating telepathic communication are unsatisfactory. In fact, the inductor can hardly experience any emotional sensations from the figures drawn on Zener cards or from the mental repetition of the instructions which he has to transmit to the percipient. There is nothing to arouse emotion in a pericipient awaiting the reception of such ordinary instructions.

* * *

Thus, the above-discussed experimental information on the generation and reception of EmFs by animals suggests that the transmission of signals between animals by methods of as yet unknown physical nature is effected by means of EmFs. Techniques for direct detection of such signal transmission between animals are outlined. At present there is no theoretical or experimental basis for assuming the existence of bioinformation transfer (or "telepathic communication") in man, since the results of parapsychological investigations obviously cannot be considered as reliable.

Chapter 14

PRACTICAL APPLICATIONS

In the light of the discussed experimental information and theoretical aspects of the biological action of EmFs and the role of EmFs in the vital activity of organisms we can indicate regions of practical application of the described effects and discuss some of the prospects in this respect.

We will not deal with the various electronic devices which have found application in medicine, biology, and agriculture. An indication of the extent of the range of such applications is the fact that for several years now there have been regular international congresses on the application of electronics in medicine and biology.

We will discuss here only some of the lines of research relating to various aspects of the biological activity of EmFs and of promise for the solution of practical problems in medicine, agriculture, biology, bionics, etc.

14.1. EmFs in Therapy and Diagnosis

EmFs of various frequency ranges have been used for some time now in physiotherapy for the treatment of various diseases. These methods of treatment entail either contact application of the EmF to the patient (galvanization, faradization, diathermy, etc.) or action at a distance (electric breeze, darsonvalization, inductothermy, UHF therapy, etc.). Mastery of SHF technique in radio engineering gave rise to a new field of physiotherapy, "microwave therapy," which rapidly found application for the treatment of several diseases. Reviews of microwave therapy (Schwan and Piersol, 1954; Alm, 1958; Presman, 1960; Presman et al, 1961; Skurikhina, 1961, 1962; and others) show that it is effective for the treatment of musculoskeletal diseases, various eye ailments, gynecological diseases, tooth diseases, etc.

However, in the introduction of the new frequency range — SHF — into physiotherapy the tendency has been to follow the procedures adopted in the case of lower-frequency EmFs. Treatment involves the use of high-intensity SHF fields, which cause considerable heating of the tissues, or medium-intensity fields ("oligo-thermic" intensities), which still cause a fair amount of heating. Yet the above-discussed results of experimental investigation of the biological action of EmFs indicate that SHF fields (and EmFs of other frequency ranges) of low nonthermal intensities might be more effective. These investigations suggest that it may be possible to direct the behavior of particular control systems in the human body. The following main trends in the development of new methods of EmF therapy appear promising.

1. The use of the "vagotonic effect" produced by whole-body exposure of man or animals to EmFs and, in particular, by direct action on peripheral receptor zones, which can quite easily be accomplished in the SHF range.

What is already known about this effect suggests that it may used, at least, for the symptomatic treatment of hypertonia of neural origin (Obrosov, 1963; Obrosov and Jasnogorodski, 1961; Obrosov et al., 1963). Further investigations will have to be directed towards determining the EmF parameters required for appropriate action on the regulation of cardiovascular function. The planning of such investigations will have to take into account the difference in the effects due to direct action of EmFs on the peripheral divisions of the nervous system and brain structures and that in the second case the effect occurs only at low intensities. Finally, the cumulative nature of the effect of low-intensity EmFs on cardiovascular regulation will have to be considered.

2. Investigations of the mechanism of the stimulating effect of EmFs on hemopoiesis and the composition of the blood in animals and man.

Effects of this kind are of particular interest in connection with radiation injury. Leucocyte production is considerably stimulated not only by subjection to an EmF before or during exposure to ionizing radiation, but also after exposure, an effect which is particularly important for possible therapeutic application. In the planning of further research in this direction it would be better to experiment with multiple exposures to a weak EmF rather than with a brief exposure to a strong field or longer exposure to a moderate field. In the choice of parameters of the acting EmFs it should be borne in mind that SHF and low-frequency fields, as well as a constant magnetic field, have been found to affect the immunological properties of the organism in the case radiation injury and infection.

3. Investigations of the inhibitory action of EmFs on malignant tumors, an effect which has been discovered only in experiments on animals.

In this case the investigations should be devoted to three main problems: First, the effect of EmFs of different frequencies and a constant magnetic field; second, the effect of the injection of extracts of normal tissues previously exposed to a EmF into patients, and third (the most promising line), the effect of combined action of UHF or SHF fields and a constant magnetic field. In all these investigations the experimental data on the genetic effects of EmFs and the results of experiments on embryos and growing cell cultures must be taken into account.

4. Investigations of the effect of spontaneous changes in natural EmFs on the dynamics of neuropsychical and cardiovascular diseases (such an effect has been discovered in clinical observations).

It would be worth while now to undertake such practical measures as the extensive organization of permanent "medicometeorological" stations to ensure the timely application of prophylactic measures in periods of magnetic storms and other spontaneous alterations of the intensity of natural EmFs. Such measures may be of a purely medical nature (bed regime, medicaments, etc.) or of a technical nature. For instance, during magnetic storms patients might be kept in screened rooms or in ordinary wards where the effect of magnetic storms is counteracted by oppositely directed artificially produced fields. It is possible that special personal devices to counteract the effect of spontaneous variations in natural EmFs might be devised in the future.

5. Further investigations of the biological effect of "magnetized" and "activated" water.

Experimental investigations of these effects have just begun, but the first results (the effect of "magnetized" water on the growth of kidney stones and on the development of animals, Piccardi's experiments with "activated" water) cannot fail to attract the attention of doctors.

These examples do not, of course, exhaust the prospects of the application of EmFs for therapeutic purposes. It should be noted that in all cases of application of EmFs to produce desirable

changes in pathological conditions there is a possible risk of disturbance of normal physiological processes in the body. In the
case of mild treatment, however, such disturbances are usually
rapidly reversible and are permissible to the same extent as similar effects due to the use of other therapeutic agents. It is obvious
that treatment will be most effective when the mechanisms of electromagnetic control systems and the nature of their functioning in
relation to natural EmFs have been determined.

In the various methods of electrodiagnosis (electrocardio-
graphy, electromyography, electroencephalography, etc.) either contact electrodes applied to appropriate regions of the body, or probes
introduced into body cavities have been used until recently. In recent years various kinds of "radioprobes" have been developed.
These are miniature radio transmitters which are introduced into
body cavities and even into arteries and give information about various chemical and physical parameters (Goldsmith, 1966).

Methods of diagnosis based on the detection of electromagnetic radiation from the organism have also been devised. An example of such a device is the above-mentioned "magnetocardiograph,"
which records the state of the heart activity from magnetic field
pulses. A method of contactless investigation of blood circulation,
based on measurement of the degree of absorption of SHF energy
in the investigated organism, has also been devised (Moskalenko,
1958, 1960). Finally, measurement of the electric parameters of
body tissues in various frequency ranges, particularly the SHF range,
can be used for diagnostic purposes (Livenson, 1964a).

Other promising lines of further research into the use of
EmFs for diagnostic purposes are as follows:

1. A study of the possibility of using the electromagnetic radiation from organs and tissues which has been detected experimentally in man and animals.

Recording of such radiation can be used for diagnosis only if
there is a detectable difference in the nature of the emission in
normal and pathological states. The demonstration of this will be
one of the first aims of research of this kind. Also of interest in
this connection is the detection (by means of contact electrodes) of
changes in the body surface potentials, which reflect the activity of
the newly discovered electromagnetic control systems (for instance,

the above-described system connected with slow transmission of information about pain and with psychic activity).

2. The development of methods of contactless measurement of the electric parameters and determination of the structure of tissues in vivo (particularly brain tissues) and contactless measurement of the activity of nervous tissues.

Measurement of the electric parameters can be based on the determination of the impedance of internal tissues from the nature of the reflection and absorption of UHF and SHF energy, and the structure of tissues can be determined from an analysis of the standing waves formed in them (Adey et al., 1962, 1963).

Contactless measurement of the activity of nervous tissues can be effected by an estimation of the concentration of free radicals in the excited nervous tissues by electron paramagnetic resonance (Kelly, 1962, 1963).

3. Comparison of the EmF absorption spectra (and dielectric parameters) of normal and malignant tissues in vitro.

Such investigations are justified by the previously mentioned differences in the resonance absorption of low-frequency EmFs by normal and malignant tissues and the corresponding differences in the dielectric parameters measured in the SHF range.

4. Investigations of the possiblity of using EmFs of various frequency ranges as a stimulating factor in functional investigations.

The results of the experiments described earlier indicate that an EmF with specially chosen parameters can be used for direct action on a particular division of the nervous system, to produce the vagotonic reaction by action on the peripheral receptor zones or the sympathicotonic reaction by action on brain structures, to enhance or suppress electric activity in various parts of the brain without the insertion of electrodes into them, to stimulate the production of leucocytes by the hemopoietic apparatus, and so on.

These are merely first outlines of possible applications of EmFs for therapy and diagnosis. The obtaining of new experimental data on the biological action of EmFs, electromagnetic control systems, and the generation of EmFs in the human body will undoubtedly lead to an extension of the field of application, but it is difficult at present to predict particular trends.

14.2 EmFs as a Hygienic Factor

The problem of occupational injury due to EmFs of various frequency ranges has been investigated fairly thoroughly, particularly by Soviet hygienists. It should be noted that the biological effect of EmFs of nonthermal intensities was first conclusively demonstrated in hygienic investigations. Maximum permissible intensities for EmFs of various frequency ranges — high to superhigh — have now been laid down and adopted in practice (Osipov, 1965; Gordon, 1966). This introduction of appropriate protective and preventive measures has practically eliminated the possibility of any harmful effect from EmFs of high and superhigh frequencies in the testing and operation of generators. Intensive research is being carried out to discover any possible occupational injury due to low-frequency EmFs and static fields.

Much less attention has been given so far to investigations of the biological action of EmFs from the viewpoint of public health. This is quite understandable. Investigations of the effect of EmFs in industrial conditions resulted from complaints of ill health (albeit slight) by persons working close to generators, but it is hardly likely that people would turn to doctors in connection with minor upsets which might be due to environmental electromagnetic fields emitted by radio and television stations.

Yet public-health workers have brought up the valid argument that if the fairly weak EmFs created in industrial conditions can have a harmful effect, then people living in the vicinity of powerful radio stations might be exposed to a similar effect. At present there is not a sufficient basis for an affirmative or negative answer to this question. This does not mean, however, that there are no grounds for raising this question. We will consider this point in the light of the experimental information discussed in the previous chapters.

As we have already mentioned (Section 2.6), the intensity of EmFs close to radio and television stations may be tenths of a volt per meter or more, and the mean level of the "radio background" is at least one or two orders greater than the level of atmospheric interference. In both cases we are dealing with intensities which are as high as those responsible for various manifestations of the biological action of EmFs. Thus, it would appear that some, even if slight, effect of the "radio background" on people is fundamentally possible. However, as distinct from industrial conditions, where people are subjected to periodic exposure to EmFs (only during part of the whole of the working day), the action of the "radio background" is practically continuous. Which is worse from the viewpoint of a possible harmful effect? It is known that the action of an external factor (for which, of course, there are compensatory reactions) for a sufficiently long time leads to permanent changes of an adaptive nature in the organism. Hence, we can surmise that the human organism has become adapted to the continuous presence of the "radio background". It is also known that adaptation to a continuously acting factor may involve an excited state of the nervous system (as, for instance, in the case of continous exposure to noise) and, hence, may lead to certain functional disturbances.

Thus, the question of the possible harmful effect of the "radio background" on man is still unresolved (as, indeed, is the question of the harmful effect of several other artificially produced factors). It is obvious that a first step here would be an appropriate statistical investigation, such as a comparison of the general state of health of people living close to powerful radio stations and people living in regions with an average "radio background."

The question of the effect of the "radio background" on the development of plants and animals is equally interesting. To what extent can these organisms adapt themselves to this continuously acting factor?

14.3. Biological Effects of EmFs in Agricultural Practice

Some of the effects produced by EmFs on plants and animals suggest that methods of acting in a desired manner on farm plants and animals will be developed in the near future. The following lines of research appear promising.

1. Experimental data show that in certain conditions EmFs may have a stimulating effect on the growth and development of crop plants. Such an effect can apparently be produced by direct exposure of seeds or seedling to EmFs of various frequencies and static fields, or by an indirect method — by treatment of the water used to water the plants. Investigations of the effect of natural EmFs on crop plants, particularly the stimulation of growth when seeds are sown in a particular orientation relative to the geomagnetic

field, are of interest. Finally, it may be possible to select the parameters of the EmF so that it has a selective inhibitory action on weeds.

2. The effect of an EmF on the reproduction of insects has
been described. In special experiments Barker et al. (1956) managed
to produce 100% mortality in grain weevils and flour beetles by
exposure to a SHF field. Further research of this kind with a view to de-
vising a new method of exterminating grain pests is of importance.

3. A recently discussed possiblity (Rait, 1964) is the "biological sterilization" of insects by saturating their habitats with the
smell of sexual attractants so that the probability of males meeting females will be reduced. If, as we believe, there is an electromagnetic link between insects of different sex, then it may be disrupted
by the production of appropriate "radio interference" or by attraction
of the males by imitated electromagnetic signals of the females.

4. Of special interest in connection with application in agriculture are the genetic effects of EmFs. The experimental data
which have been obtained indicate that exposure of animals and
plants to EmFs with particular parameters can be used to obtain
desirable mutations, alter the sex ratio in the progeny, and so on.

14.4. EmFs and Space Biology

In connection with various problems of space biology some
foreign scientists (Beischer, 1962, 1965; Alexander, 1962;
R. Becker, 1963; Goldsmith, 1966; and others) have recently raised
questions connected with the biological action of EmFs.

So far space flights by man, animals, and plants have taken
place at relatively short distances from the earth, where the conditions as regards natural EmFs are only slightly altered. What
will be the situation, however, in long space trips, where the crew
and all the living things in the space ship will be deprived of the
effect of terrestrial EmFs? Will vital processes in the "micro-
biosphere" of the space ship still be normal in these conditions?
First observations on subjects kept for ten days in conditions where the
earth's magnetic field was compensated (Beischer, 1965) did not show
any appreciable physiological changes, but there was some disturbance
of visual perception — the flicker fusion threshold was reduced.

Questions of the opposite nature also arise: How will the

living population of a space ship be affected by the strong magnetic fields which are to provide protection from cosmic rays? How will living organisms react to weak magnetic fields of cosmic origin with a periodicity different from that of the geomagnetic field?

In the consideration of conditions in space we cannot ignore such features as the absence of the earth's electric field and the EmF of atmospherics in space ships.

In the light of the above-described experimental data on the effect of natural EmFs on living organisms all these questions cannot be regarded a priori as of no account. The very fact that these questions have been raised in foreign journals on astronautics and space biology indicates that they merit consideration.

14.5. Use of EmFs in Biological Investigations

EmFs have already found wide application in various biological investigations. We need only mention the radiospectroscopy of biological objects, particularly by the techniques of electron paramagnetic and nuclear magnetic resonance. We will give some examples of recently developed biological research techniques employing EmFs.

Several investigations (Haggis et al., 1951; Buchanan et al., 1952; Grant, 1957) determined the degree of hydration of protein molecules from the dispersion of the dielectric constant in the SHF range. This method is fairly simple but it is not very accurate. Measurements of the electric conductivity of intracellular contents in the SHF range (Cook, 1952b) are much more accurate than measurements made by previous methods. Methods of measuring the electric parameters of crystalline proteins, amino acids, and peptides in a wide frequency range — from 1000 Hz to 4000 MHz — have been devised (Bayley, 1951). Special methods of measuring the complex dielectric constant of thin films and fibers in the SHF range (Shaw and Windle, 1950) allow a qualitative and quantitative assessment of the absorption of water on wool fibers (Windle and Shaw, 1954, 1956).

Prospects of using an EmF as a "tool" in biological research are indicated by the results of the above-discussed experimental investigations of the biological action of EmFs.

1. Detection of the EmFs generated in different parts of the entire organism is of interest not only as an index of the electromagnetic processes associated with vital activity, but also as a test in attempts to discover the mechanisms of physiological processes. On the other hand, EmFs can be used to act locally on particular structures of the organism without gross interference with the processes under investigation. Of equal interest is the use of EmFs in the investigation of conditioned-reflex activity, since this factor can act as a conditioned stimulus without causing an appreciable reaction of the known sense organs. Finally, exposure of organisms to electromagnetic fields with parameters close to those characteristic of bioelectric activity can provide additional information about the role of this activity in physiological processes.

2. In the investigation of cultures of cells and unicellular organisms EmFs of appropriate frequencies and intensities can be used either as an inhibiting or stimulating factor. Measurement of the electric properties of cell suspensions in different frequency ranges may be an effective method of determining the nature of cell structures (as some investigations of such kind have shown). The possibility of using EmFs to act on intracellular systems, particularly the genetic apparatus, is of great interest. Such interference may produce changes in intracellular structures (chromosome aberrations, for instance) in a medium of normal chemical composition and in normal physical conditions.

3. The use of EmFs in experiments at the molecular level may be very rewarding. Investigations of the change in optical properties of protein solutions due to EmFs, the resonance absorption of EmFs by molecules in solutions and in the crystalline state, and of dielectric properties of protein molecules in various frequency ranges may contribute to a more thorough analysis of the structure of macromolecules. Investigations of the electromagnetic interconnection between macromolecules may reveal the mechanism of such interactions as that of the enzyme—substrate reaction, the reaction between DNA and RNA, and so on.

14..6. Biological Activity of EmFs

and Bionics

According to McCulloch's definition (1962), bionics is concerned mainly with the investigation of the methods which nature

has resorted to for the solution of various problems and the ultimate aim of bionics is to embody these methods in instruments and
devices. To bionics is assigned one of the main roles in biology:
the study of the structure and functions of biological systems, the
nature of various biological processes, and the mechanisms of the
interactions responsible for the existence and development of living organisms. Is it reasonable to include all these problems in
bionics? It appears to be reasonable since their solution requires
the extensive collaboration of representatives of various scientific
disciplines — biologists and doctors, biophysicists and biochemists,
physicists and mathematicians, cyberneticists and engineers, and
such a situation is created around bionics.

If this is accepted as the sphere of bionics this branch of
science is faced with the solution of problems connected with the
determination of the mechanisms of the biological activity of EmFs
and the establishment of the physicobiological bases of the role of
EmFs in the vital activity of organisms. We will briefly list some
problems of this kind.

1. The investigation of the design of the "multistage" electromagnetic regulation in living organisms, the determination of
the structure and function of the biological systems implicated in
such regulation, the evaluation of their "working parameters," and
so on.

We stress once again that such investigations must begin with
the entire organism in its natural conditions of existence. It is only
after the general scheme of regulation in the organism has been
established and the "hierarchy" of this regulation has been determined that individual isolated systems and their constituent parts
can be investigated.

2. The investigation of the mechanism of the regulating and
disturbing effect of environment EmFs on the regulation of vital
processes of organisms and their development whereby the best
interrelationship of organism and environment is secured.

It will be of particular interest to discover the mechanisms
of the very high and selective sensitivity of living organisms to
EmFs and the great resistance to interference of biological systems
due to the presence of multistage passive and active shielding.
The main parameters of these mechanisms might be evaluated by

investigating the behavior of organisms when they are exposed to natural EmFs and when these EmFs are weakened, distorted, or eliminated. It will then be worthwhile to carry out laboratory investigations (with artificially produced appropriate EmFs) directed towards the discovery of the principles of design and function of the biological systems which receive and convert external electromagnetic information.

3. The investigation of the biological systems responsible for the orientation and navigation of living organisms in all their habitats.

The approach to this problem may be based not only on the postulated existence of special systems of orientation and navigation in individuals, but also from the standpoint of the above-mentioned idea of communal systems formed in communities of organisms. First, it should be borne in mind that the orientational guides for animals may be not only the earth's magnetic and electric fields, but also "radio beacons" — other individuals emitting radiation.

4. The investigation of electromagnetic interconnections in animal communities, which are regarded as self-organizing systems, is of considerable interest. As we saw, in several cases the community emerges as a "mediator" between the individual and the environment and effects regulation of the common behavior of individuals and the size of the population in accordance with the changes in the environment.

The solution of this problem, however, depends not so much on the correct design of experimental investigations, as on the development of a system of general mathematical logic which can be used to describe the diverse interactions in an ensemble consisting in turn of complex members. Attempts to create such a logic are being undertaken by some scientists (see, for instance, the cited paper of Foerster).

Such in general are the problems which may be included in the sphere of bionics. As regards the aim of bionics, the design of instruments, the results of investigations of the biological activity of EmFs can, of course, provide information which can be useful for the solution of practical problems of electronics, the design of electromagnetic control systems, and so on.

CONCLUSION

The aim of this book was to pose the biological problem of the significant role in living nature of electromagnetic fields of infralow to superhigh frequencies, i.e., the extensive region of the electromagnetic spectrum which was previously regarded as of no biological significance. The formulation of this problem is based on the hypothesis of informational functions of EmFs in the vital activity of organisms. These functions are manifested in three ways — in the transmission of information from the environment to organisms, in informational interconnections with organisms, and in the exchange of information between organisms.

We have tried to substantiate the formulation of the problem and the basic hypothesis, and to suggest ways in which the latter might be experimentally verified, by approaches from various angles: first from a generalization and analysis of the collected experimental information on the biological effects of EmFs; second from a consideration of the general features manifested in these effects; and third in the light of general biological considerations of the possible role of EmFs in life.

We will try now to assess how far we have succeeded in this substantiation — what aspects of the problem are adequately supported by experimental data and what aspects are based only on indirect evidence and hypotheses of varying feasibility.

First of all, we think that there are sufficient grounds for asserting that the diverse manifestations of the biological action of EmFs indicate that living organisms have specific properties formed in the process of evolutionary development. It is only on the basis of this hypothesis that we can explain the experimentally observed high sensitivity to EmFs of organisms of the whole evolutionary hierarchy (from unicellular organisms to man), the reactions of the most diverse biological systems (from the molecular to the organismic level) and, finally, the sensitivity of living systems to changes in natural environmental EmFs.

An analysis of the relationship between the biological effects of EmFs and their parameters and site of action corroborates the hypothesis of the informational (and not energetic) nature of the interaction of EmFs with biological systems. The regulating effect of natural EmFs on living organisms and the disturbance of regulation and interconnections in the organism due to inappropriate artificial EmFs of various parameters have been observed at intensities at which any appreciable energetic effects in the tissues are quite impossible. Moreover, according to the available information, the nature of the reactions of organisms to EmFs depends not on the amount of electromagnetic energy absorbed in the tissues, but mainly on the modulation and time parameters of the EmF and on which particular systems of the organism are acted on. In addition, the magnitude of a particular reaction is not proportional to the intensity of the acting EmFs but, on the contrary, decreases with increase in intensity in several cases. Some reactions produced by the action of weak EmFs are not produced at all at high intensities.

Experimental observations indicate fairly definitely that living organisms can orient themselves relative to the earth's electric and magnetic fields. The disturbance of physiological functions (especially in pathological states of organisms) due to sporadically varying natural EmFs can also be regarded as an established fact. It is very probable that periodically varying environmental EmFs affect the rhythms of physiological processes in a wide variety of organisms. However, there will have to be many more investigations to separate this effect from that of other periodically varying geophysical factors.

There is still no direct evidence of the existence of EmF interconnections in the considered range of frequencies within the organism. However, the results of electrophysiological investigations in the infralow- and low-frequency ranges and the data indicating the generation and reception of EmFs of various frequencies in biological systems suggest that such evidence will be obtained in the near future.

The possibility of EmF interconnections between organisms is indicated at present only by indirect evidence: It has been es-

tablished that living organisms create electromagnetic fields of
various frequencies in the space surrounding them and that they
are highly sensitive to these fields.

On the basis of the concept of the informational functions of
EmFs in living nature we have expressed several hypotheses on
these interactions which are effected within the organism and between organisms by mechanisms of still unkown physical nature.
We have also suggested some experimental methods of verifying
these hypotheses. Finally, we have expressed our opinion on the
general approach to the investigation of the biological effects of
EmFs. We think that 1) research must begin with investigations
on entire organisms (and, in some cases, on groups and communities of organisms) and must then be followed by a gradual descent
down the hierarchy of systems and 2) simulation of these effects
(and the biological systems implicated in them) must be based on
the consideration of equivalent "organized" electromagnetic circuits.

Many of the views expressed in this book definitely lack a
sufficiently sound basis and some of them may well be erroneous,
but in this respect it is worthwhile to recall the words of Charles
Darwin: "Incorrect facts are extremely harmful to the progress
of science, since they often prevail for a very long time. Incorrect
theories, however, if they are supported by any data at all, do not
cause harm, since each one finds a saving pleasure in their refu-
tation, and when their erroneousness is demonstrated, then one of
the pathways to error is closed and at the same time the way to
the truth is opened."

In conclusion we can say that the problem of the informational
role of electromagnetic fields in life clearly merits thorough discussion and the conduction of extensive experimental investigations.
If this book attracts the interests of the readers to this problem
and, perhaps, increases the number of researchers in this field,
the author will have his highest reward.

Reichenbach's Researchs, 1850, hL 162/3

REFERENCES

Abros'kin, V. V. (1966). Some results of the action of the earth's magnetic field on plants, in: Proceedings of Conference on the Effect of Magnetic Fields on Biological Objects, Moscow, p. 11.

Addington, C., et al. (1961). Biological effects of microwave energy at 200 mc, in: Biological Effects of Microwave Radiation, Vol. I, Plenum Press, New York, p. 177.

Adey, W., et al (1962). Impedance measurement in brain tissues of animals using microvolt signals, Exptl. Neurol., 5: 47.

Adey, W., et al. (1963). Impedance changes in cerebral tissue accompanying a learned discriminative performance in the cat, Exptl. Neurol., 7: 259.

Aghina, C., et al. (1965). Il tenore in acidi grassi a catena lunga e numero pari di atomi di carbonio (da C^{16} a C^{18}) della quota lipidica del latte prodotto da bovine di razza frisona n. p. sottoposte, a livello mammario, all'azione di un campo magnetico rotante in bassa frequenza, Atti Soc. Ital. Sci. Veter., Vol. 9.

Akhmerov, U. Sh., et al. (1966). The effect of a constant magnetic field on the developmental phases of Drosophila, in: Proceedings of Conference on the Effect of Magnetic Fields on Biological Objects, Moscow, p. 10.

Akoyunoglou, G. (1964). Effect of a magnetic field on carboxydismutase, Nature, 202: 425.

Aladzhalova, N. A. (1962). Slow Electric Processes in the Brain, Izd. Akad. Nauk. SSSR, Moscow.

Aleksandrovskaya, M. M., and Kholodov, Yu. A. (1965). The reaction of brain neuroglia to exposure to a constant magnetic field, in: Questions of Hematology, Radiobiology, and the Biological Action of Magnetic Fields, Tomsk, p. 342.

Aleksandrovskaya, M. M., and Kholodov, Yu. A. (1966). Reactions of cat brain neuroglia to exposure to a constant magnetic field, in: Proceedings of Conference on the Effect of Magnetic Fields on Biological Objects, Moscow, p. 3.

Alexander, H. (1962). Biomagnetics: the biological effects of magnetic fields, Am. J. Med. Electronics, 1: 181.

Alm, H. (1958). Einführung in die Microwellen-Therapie, Berliner Med. Verlag, Berlin.

Alvarez, A. (1935). Apparent points of contact between the daily course of the magnetic components of the earth together with certain solar elements and the diastolic pressure of human beings and the total count of their leucocytes, Puerto Rico J. Public Health, 10: 734.

Amer, N. (1956). An observation on the detection by the ear of microwave signals, Proc. IRE, 44: 2A.

Amer, N., and Tobias, C. (1965). Analysis of the combined effect of magnetic fields, temperature, and radiation on development, Radiation Res., 25: 172.

Amineev, G. A., and Sitkin, M. I. (1965). The effect of a low-frequency alternating magnetic field on the behavior of mice in a T-shaped maze, in: Questions of Hematology, Radiobiology, and the Biological Action of Magnetic Fields, Tomsk, p. 372.

Amineev, G. A., and Khasanova, R. I. (1960). The effect of a constant magnetic field on the myoneural synapse, in: Proceedings of Conference on the Effect of Magnetic Fields on Biological Objects, Moscow, p. 7.

Anne, A., et al. (1961). Relative microwave absorption cross sections of biological significance, in: Biological Effects of Microwave Radiation, Vol. 1, Plenum Press, New York, p. 153.

Antonov, I. V., and Plekhanov, G. F. (1960). A possible mechanism of the primary action of a magnetic field on the elements of living systems, in: Materials of Theoretical and Clinical Medicine, No. 2, Tomsk, p. 127.

Aschoff, J. (1960). Exogenous and endogenous components in circadian rhythms, Cold Spring Harbor Symp. Quant. Biol. 25: 11.

Audus, L. (1960). Magnetotropism: a new plant-growth response, Nature, 185: 132.

Audus, L., and Whish, J. (1964). Magnetotropism, in: Biological Effects of Magnetic Fields, Plenum Press, New York, p. 170.

Austin, G., and Horwath, S. (1949). Production of convulsions in rats by exposure to ultrahigh-frequency electrical currents (radar), Am. J. Med. Sci., 218: 115.

Austin, G., and Horwath, S. (1954). Production of convulsions in rats by high-frequency electrical currents, Am. J. Phys. Med., 33: 141.

Averkiev, M. S. (1960). Meteorology, Izd. MGU.

Bach, S. (1961). Changes in macromolecules produced by alternating electric fields, Digest Internat. Conf. Med. Electronics, 21: 1.

Bach, S. (1965). Biological sensitivity to radiofrequency and microwave energy, Federation Proc., 24 (pt. 3): 22.

Bach, S., et al. (1961a). Effects of R-F energy on human gamma globulin, J. Med. Electronics, Vol. 9, Sept.-Nov.

Bach, S., et al. (1961b). Effect of radio-frequency energy on human gamma globulin, in: Biological Effects of Microwave Radiation, Vol. 1, Plenum Press, New York, p. 117.

Baldwin, M., et al. (1960). Effect of radio-frequency energy on primate cerebral activity, Neurology, 10: 178.

Baranski, S., et al. (1963). Recherches expérimentales sur l'effet mortel de irradiation des ondes micrométriques, Rev. Med. Aeron., 2: 108.

Barber, D. (1961). The reaction of luminous bacteria to microwave radiation in the frequency range of 2608.7-3082.3 MC, Techn. Rept. Univ. Michigan School of Public Health, April.

Barker, V., et al. (1956). Some effects of microwaves on certain insects which infect wheat and flour, J. Econ. Entomol., 49:33..

Barlow, H., et al. (1947). Visual sensations aroused by magnetic fields, Am. J. Physiol., 148: 372.

Barnothy, J. (1963). Growth rate of mice in static magnetic fields, Nature, 200: 86.

Barnothy, J. (1964). Basic concepts related to magnetic fields and magnetic susceptibility, in: Biological Effects of Magnetic Fields, Vol. I, Plenum Press, New York, p. 3.

Barnothy, M. (1963a). Reduction of radiation mortality through magnetic pretreatment, Nature, 200: 279.

Barnothy, M. (1963b). Biological effects of magnetic fields on small mammals, in: Biomedical Sciences Instrumentation, Vol. 1, Plenum Press, New York, p. 127.

Barnothy, M. (1964). Development of young mice, in: Biological Effects of Magnetic Fields, Vol. I, Plenum Press, New York, p. 93.

Barnothy, M., and Barnothy, J. (1960). Biological effects of magnet-
    ic fields, in: Medical Physics, Vol. 3, The Year Book Publ.,
    Chicago, p. 61.
Barnothy, M., and Barnothy, J. (1963). Second-day minimum in the
    growth curve of mice subjected to magnetic fields, Nature,
    200: 189.
Barnwell, F., and Brown, F. (1964). Responses of planarians and
    snails, in: Biological Effects of Magnetic Fields, Vol. 1,
    Plenum Press, New York, p.263.
Bartoncek, V., and Klimkova-Deutscheva, E. (1964).Casopis Lekaru
    Ceskych, 1: 26.
Baule, G., and McFee, R. (1963). Detection of the magnetic field
    of the heart, Am. Heart J., 66:95.
Bayley, S. (1951). The dielectric properties of various solid crys-
    talline proteins, amino acids, and peptides, Trans. Faraday
    Soc., 47: 509.
Becker, G. (1963a). Magnetfeld-orientierung von Dipteren; Natur-
    wissenschaften, 50: 564.
Becker, G. (1963b). Reaction of insects to magnetic fields, in: Se-
    cond Internat. Biomagnet. Symp., p. 19.
Becker, R. (1963a). The biological effects of magnetic fields. A
    survey, Med. Electronics and Biol. Engng, 1: 293.
Becker, R. (1963b). Relationship of geomagnetic environments to
    human biology, N. Y. State J. Med., 63: 2215.
Becker, R., et al. (1962). The direct current control system. A
    link between environment and organism, N. Y. State J. Med.,
    62: 1169.
Beischer, D. (1962). Human tolerance to magnetic fields, Astronau-
    tics, 7: 46.
Beischer, D. (1964). Survival of animals in magnetic fields of
    140,000 Oe, in: Biological Effects of Magnetic Fields, Vol.I,
    Plenum Press, New York, p. 201.
Beischer, D. (1965). Biomagnetics, Ann. N. Y. Acad. Sci., 134:
    454.
Belitskii, B. M., and Knorre, K. G. (1960). Protection from radia-
    tions in work with SHF generators, in: The Biological Action
    of Superhigh Frequencies, Moscow, p. 107.
Belova, S. F., and Gordon, Z. V. (1956). The effect of centimeter
    waves on the eyes, Byull. Eksperim. Biol. i Med., No. 4: 43.
Belova, S. F.(1960a). The state of the visual organ in persons ex-

posed to superhigh frequency fields, in: Physical Factors of
the External Environment, Moscow, p. 184.

Belova, S. F. (1960b). Change in the elastotonometric curve in rabbits after exposure to SHF, in: The Biological Action of Superhigh Frequencies, Moscow, p. 86.

Bennett, M. (1964). The mechanism of the action of electric organs,
in: Modern Problems of Electrobiology [Russian translation],
Mir, Moscow.

Berg, H. (1960). Phénomènes solaires et terrestres en biologie,
in: Relations entre phénomènes solaires et terrestres en
chimie-physique et biologie, Press Acad. Europ., Brussels,
p. 160.

Bibbero, R. (1951). Telepathic communication, Proc. IRE, 39: 290.

Biotelemetry (1965). Mir, Moscow.

Blois, M. S. (ed.) (1961). Free Radicals in Biological Systems,
Academic Press, New York.

Blyumenfel'd, L. A., et al. (1962). The Application of Electron Paramagnetic Resonance Chemistry, Izd. Sibirsk. Otdel. Akad. Nauk
SSSR, Novosibirsk.

Bodrova, N. V., and Krayukhin, B. V. (1965). The lateral line of
fish as an apparatus for the perception of an electric field,
in: Bionics, Nauka, Moscow, p. 264.

Boe, A., and Salunkhe, D. (1963). Effects of a magnetic field on
tomato ripening, Nature, 199: 91.

Boyle, A., et al. (1950). The effects of microwaves. Preliminary
investigation, Brit. J. Phys. Med., 13: 2.

Boysen, J. (1953). Hyperthermic and pathologic effects of electromagnetic radiation (350 Mc), Arch. Industr. Hyg., 7: 516.

Brady. J. (1958). The paleocortex and behavioral motivation, in:
Biological and Biochemical Bases of Behavior, H. F. Harlow
and C. N. Woolsey (eds.), University of Wisconsin Press,
Madison.

Brandt, A. A. (1963). Research on Dielectrics at Superhigh Frequencies, Gos. Izd. Fiziko-Matem. Lit., Moscow.

Brazier, M. (1961). The Electrical Activity of the Nervous System,
Macmillan, New York.

Brown, F. (1959). Living clocks, Science, 130: 1535.

Brown, F. (1960). Response to pervasive geophysical factors and
the biological clock problem, Cold Spring Harbor Sym. Quant.
Biol. 25: 57.

Brown, F. (1962a). Extrinsic rhythmicality: A reference frame
    for biological rhythms under so-called constant conditions,
    Ann. N. Y. Acad. Sci., 98: 775.
Brown, F. (1962b). Responses of the planarian Dugesia and the
    protozoan Paramecium to very weak horizontal magnetic
    fields, Biol. Bull., 123: 264.
Brown, F. (1962c). Response of the planarian Dugesia to very weak
    horizontal electrostatic fields, Biol. Bull., 123:282.
Brown, F. (1963a). The biological rhythm and its bearing on space
    biology, in: Bioastronautics — Fundamental and Practical
    Problems, Vol. 17, p. 29.
Brown, F. (1963b). How animals respond to magnetism, Discovery,
    November.
Brown, F. (1963c). An orientational response to weak gamma ra-
    diation, Biol. Bull., 125: 206.
Brown, F., et al. (1956). Solar and lunar rhythmicity in the rat in
    "constant conditions" and mechanism of physiological time
    measurement, Am. J. Physiol., 184: 491.
Brown, F., et al. (1960a). A magnetic compass response of an or-
    ganism, Biol. Bull., 118: 382.
Brown, F., et al. (1960b). Magnetic response of an organism and
    its solar relationships, Biol. Bull., 118: 367.
Brown, F., et al. (1964a). Adaptation of the magnetoreceptive mech-
    anism of mud-snails to geomagnetic strength, Biol. Bull.,
    127: 221.
Brown, F., et al. (1964b). A compass directional phenomenon in
    mud-snails and its relation to magnetism, Biol. Bull., 127:
    206.
Brown, F. A., Jr., and Park, Y. H. (1964). Seasonal variations in
    sign and strength of gamma-taxis in planarians, Nature, 202:
    469.
Brown, F. A., Jr., and Park, Y. H. (1965). Duration of an after-
    effect in planarians following a reversed horizontal magnetic
    vector, Biol. Bull., 128: 347.
Brown, F. A., Jr., Park, Y. H., and Zeno, J. R. (1966). Diurnal
    variation in organismic response to very weak gamma radia-
    tion, Nature, 211: 830.
Brown, F. A., Jr. and Park, Y. H. (1967). Association-formation
    between photic and subtle geophysical stimulus patterns. —
    A new biological concept, Biol. Bull., 132: 311.

Brown, G., and Morrison, W. (1954). An exploration of the effects of strong radio-frequency fields on microorganisms in aqueous solutions, Food Technol., 8: 361.

Brown, G., and Morrison, N. (1956). An exploration of the effects of strong radio-frequency fields on microorganisms in aqueous solution, Trans. IRE Med. Electronics, PGME, 4: 16.

Bruns, S. A., et al. (1966). Change in light extinction by water after exposure to a magnetic field, Kolloidn. Zh., 28: 153.

Buchanan, T. (1952). Balance methods for measurement of permittivity in microwave region, Proc. IRE, 99 (Pt. 3): 61.

Buchanan, T., et al. (1952). Dielectric estimation of protein hydration, Proc. Roy. Soc. (London), 213A: 379.

Budko, L. N. (1964a). Change in blood carbohydrate content due to the action of electromagnetic vibrations of audio- and radio-frequency ranges on organisms, in: Some Questions of Physiology and Biophysics, Voronezh, p. 73.

Budko, L. N. (1964b). Dynamics of carbohydrate metabolism in isolated liver of white rats on exposure to electromagnetic fields of different frequencies, in: Some Questions of Physiology and Biophysics, Voronezh, p. 31.

Bullock, T. (1959). Initiation of nerve impulses in receptor and central neurons, Rev. Mod. Phys. 31: 504.

Bunning, E. (1964). The Physiological Clock, Springer-Verlag, Berlin.

Burr, H., and Mauro, A. (1949). Electrostatic fields of the sciatic nerve in the frog, Yale J. Biol. Med., 21:455.

Butler, B., and Dean, W. (1964). The inhibitory effect of a magnetostatic field upon the tissue culture of K. B. cells, Am. J. Med. Electronics, 3:123.

Bychkov, M. S., and Moreva, Z. E. (1960). The effect of radiowaves in the SHF range on a frog nerve–muscle preparation, Tr. Leningr. Obshchest. Usp'tatel. Prirod., 71:178.

Caramello, E., et al. (1965). La dinamica delle variazion quantitative di alcuni acidi grassi (C^{12} and C^{10}) e del loro rapporto nel latte vaccino: indagini sulla quota lipidica nel prodotto secreto dalla ghiandola mammaria sottoposta agli effeti di un campo magnetico rotante in bassa frequenza, Ann. Fac. Med. Vet. Torino, Vol. 15.

Carpenter, R., et al. (1960). Opacities in the lens of the eye experimentally induced by exposure to microwave radiation, Trans. IRE. Med. Electronics, ME-7:152.

Carpenter, R., et al. (1961). The effect on the rabbit eye of micro-wave radiation at X-band regions, Digest Internat. Conf. Med. Electronics, 26:5.

Cassiano, O., et al. (1966). Alcune riposte vasomotorie dell'uomo sottoposto all'azione di un campo magnetico, Minerva Anest., 32:32.

Cater, D., et al. (1964). Combined therapy with 220-kv roentgen and 10-cm microwave heating in rat hepatoma, Acta Radiol., 2:321.

Cazzamali, F. (1925). Phénomènes télépsychiques et radiations cérébrales, Neurologica, 6:193.

Cazzamali, F. (1928). Les ondes électromagnétiques correlatés avec certains phénomènes psychosensoriels, Compt. Rend. III Congr. Internat. Rech. Psychiques, Paris.

Cazzamali, F. (1929). Experienze agromenti et problemi de bio-fisica cerebrale, Quaderni di Psichiatrica, 16:81.

Cazzamali, F. (1935). Di un fenomeno radiante cerebropsichico come mezzo di esporazione psichobiofisica, Giorn. Psichia-tria, Neuropatol., 63:45.

Cazzamali, F. (1936). Phénomènes électromagnètiques du cerveau humain en active psychosensorielle intense et leir démonstra-tion par des complexes oscillateurs révélateurs à triodes pour ondes ultra-courtes, Arch. Neurol., 54:113.

Cazzamali, F. (1941). Di nuovo apparato radio-electro rivelatore dei fenomeni electromagnetici radianto del cervello umano, L'Energio Electrica, 18:28.

Cellino Tosi, A., et al. (1965). Variazoni indotte da un campo mag-netico rotante in bassa frequenza applicato alla ghiandola mammaria sul tasso di acido linoleicoe di acido linoleico nella quota lipidica del latte secreto da bovine di razza frisona n.p., Ann. Fac. Med. Vet. Torino. Vol. 15.

Chalazonitis, N. and Arvanitaki, A. (1965). Effets du champ mag-nétique constant sur l'autoactivité des fibres myocardiques, Compt. Rend. Soc. Biol., 10:1962.

Chauvin, R. (1965). From the Bee to the Gorilla [Russian transla-tion], Mir, Moscow.

Chechik, P. O. (1953). Radio-Engineering and Electronics in Astro-nomy, Gosénergoizdat, Moscow.

Chernavskii, D. S., et al. (1967). Elastic deformations of proteins, in: Proceedings of Symposium on Physics and Biology, Moscow, p. 4

Chernyshev, V. G. (1966). The effect of disturbances of the earth's magnetic field on insect activity, in: Proceedings of Conference on the Effect of Magnetic Fields on Biological Objects, Moscow, p. 80.

Chirkov, M. M. (1964). The effect of the energy of electromagnetic vibrations of the acoustic spectrum on catalase activity of blood, in: Some Questions of Physiology and Biophysics, Voronezh, p. 25.

Chirkov, M. M. (1965). Author's Abstract of Candidate's Dissertation: The Effect of the Energy of Electromagnetic Vibrations of the Audio- and Radio-frequency Ranges on the Catalase and Peroxidase Activity of Rabbit Blood and Rat Tissues, Voronezh.

Chizhenkova, R. A. (1966). Author's Abstract of Candidate's Dissertation: An Investigation of the Role of Specific and Nonspecific Formations in the Electric Reactions of the Rabbit Brain to UHF and SHF Electromagnetic Fields and a Constant Magnetic Field, Moscow.

Chizhevskii, A. L. (1963). The Sun and Us, Znanie, Moscow.

Chizhevskii, A. L. (1964). One form of the specifically bioactive or Z-emission of the sun, in: Life in the Universe, Mysl', Moscow, p. 342.

Chubaev, P. P. (1966). The effect of plant growth stimulators and inhibitors and lunar phases on the orientation of seed embryos in the earth's magnetic field, in: Proceedings of Conference on the Effect of Magnetic Fields on Biological Objects, Moscow, p. 82.

Ciba Foundation Symposium on Extrasensory Perception (1956). J. Churchill, London.

Ciecura, L., and Minecki, L. (1964). Rozmieszenie i aktywnosc niektorych enzymow hydrolitycznych w jadrach szczurow poddawanych dzialania, Med. Pracy, 15:159.

Clark, J. (1950). Effects of intense microwave radiation on living organisms, Proc. IRE, 38:1028.

Cole, K. (1950). Ionic electrical conduction of nerves, in: Regulation Processes in Biology, Oldenbourg, Munich.

Cole, K., and Cole, R. (1941). Dispersion and absorption in dielectrics, J. Chem. Phys., 9:34.

Commoner, B. (1962). Is DNA a self-reproducing molecule, in: Horizons in Biochemistry, Academic Press, New York.

Cook, E., et. al. (1964). Increase of trypsin activity, in: Biological Effects of Magnetic Fields, Vol. 1, Plenum Press, New York, p. 246.

Cook, H. (1951a). The dielectric behavior of some types of human tissues at microwave frequencies, Brit. J. Appl. Phys., 2:295.

Cook, H. (1951b). "Dielectric behavior of human blood at microwave frequencies, Nature, 168: 247.

Cook, H. (1952a). A physical investigation of heat production in human tissues when exposed to microwave, Brit. J. Appl. Phys., 3:1.

Cook, H. (1952b). A comparison of the dielectric behavior of pure water and human blood at microwave frequencies, Brit. J. Appl. Phys., 3:249.

Cook, H. (1952c). The pain threshold for microwave and infrared radiations, J. Physiol. (London), 118:1.

Cook, H., and Buchanan, T. (1950). Dielectric behaviour of methyl palmitate at microwave regions, Nature, 165:358.

Cooper, R. (1946). The electrical properties of salt-water solutions over the frequency range 1-4000 Mc, J. Inst. Elec. Engrs., 93(Pt. 3):69.

Course in Electrical Measurements (1960). Moscow.

Daily, L., et al. (1951). Influence of microwaves on certain enzyme systems in the lens of the eye, Am. J. Ophthalmol,. 34:1301.

Dainotto, F., et al. (1962). Studio del frazioni glicidiche nel musculo schelectrico dell'animale da esperimento trattato con microonde, Policlinico, 69:270.

Danilevskii, V. Ya. (1901). Investigations of the Physiological Action of Electricity at a Distance, Kharkov.

Danilewski, V. (Danilevskii, V.) (1905). Beobachtungen über eine subjective Lichtempfindung im variablen magnetischen Felde,, Pflügers Arch. Ges. Physiol. 108:513.

Dardymov, I. V., et al. (1965). The effect of water treated with a magnetic field on plant growth, in: Questions of Hematology, Radiobiology, and the Biological Action of Magnetic Fields, Tomsk, p. 325.

Dardymov, I. V., et al. (1966). The effect of water treated with a magnetic field on biological objects, in: Proceedings of Conference on the Effect of Magnetic Fields on Biological Objects, Moscow, p. 25.

D'Arsonval, A. (1893) Production des courants de haute fréquence et de grande intensité, leurs effets physiologiques, Compt. Rend. Soc. Biol. 45:122.

Deichmann, W., et al. (1959). Acute effects of microwave radiation on experimental animals (2400 Mc), J. Occup. Med., 1:369.

Deichmann, W. (1961). Factors that influence the biological effects of microwave radiation, Ind. Med. Surg., 30:264.

Deichmann, W., et al. (1961). Microwave radiation of 10 mW/cm^2 and factors that influence biological effects at various power densities, Ind. Med. Surg., 30:221.

Deichmann, W., et al. (1964). Effect of microwave radiation on the hemopoietic system of the rat, Toxicol. Appl. Pharmacol., 6:71.

Dement'ev, G. P. (1965). Questions of bionics in ornithological investigations, in: The Migration of Birds and Mammals, Nauka, Moscow, p. 11.

Dijkgraaf, S. (1964). Electroreception and ampullae of Lorenzini in Elasmobranchs, Nature, 201:523.

Dolatkowski, A., et al. (1963) Badania nad wplywem mikrofal apparatury radarowej na jadra i najadrza krolikow, Polski Przegl. Chir., 35:1221.

Dolatkowski, A., et al. (1964). Studies on the effect of microwaves emitted by radar devices on the testicles and epididymides of the rabbit, Polish Med. J., 3:1156.

Dolina, L. A. (1961). Morphological changes in the central nervous system due to the action of centimeter waves on the organism, Arkh. Patol., 1:51.

Dorfman, Ya. G. (1962). The specific nature of the action of magnetic fields on diamagnetic macromolecules in solution, Biofizika, 7:733.

Dorfman, Ya. G. (1966). The Physical Mechanism of the Action of Static Magnetic Fields on Living Systems, Izd. VINITI, Moscow.

Drischel, H. (1956). Blood sugar regulation, in: Regulation Processes in Biology, Oldenbourg, Munich.

Drogichina, É. A., et al. (1962). Some clinical manifestations of chronic exposure to centimeter waves, Gigiena Truda i Prof. Zabolwaniya, No. 1:28.

Dryden, J., and Jackson, W. (1948). Dielectric behaviour of methyl palmitate: evidence of resonance absorption, Nature, 162:676.

Dryl, S. (1965). Electrified animals, New Scient., 26:303.

Duhamel, J. (1958). Dangers biologiques des ondes courtes, Presse Med., 66:744.

Duhamel, J. (1959). Effets biologiques des ondes radioélectriques ultracourtes, Presse Med., 67:151.

Dull T., and Dull, B. (1935). Zusammenhänge zwischen Störungen des Erdmagnetismus und Häufungen von Todësfallen, Deut. Med. Wochschr., 61:95.

Eakin, S., and Thomson, R. (1962)- Effects of microwave radiation on activity level of rats, Psychol. Rept., 11:192.

Edwards, D. (1960a). Effects of artificially produced atmospheric electrical fields upon the activity of some adult Diptera, Can. J. Zool., 38:899.

Edwards, D. (1960b). Effects of experimentally altered unipolar aeroion density upon the amount of activity of the blowfly *Calliphora vicina* R. D., Can. J. Zool., 38:1079.

Edwards, D. (1961). Influence of an electrical field on pupation and oviposition in *Nepytia phantasmaria* Stkr., Nature, 191:976.

Egan, W. (1957). Eye protection in radar fields, Elec. Eng. 76:126.

Ehrlich, P., and Holm, R. (1963). The Process of Evolution, Mc-Graw-Hill, New York.

Él'darov, A. L. (1965). The effect of a constant magnetic field on birds, in: Questions of Hematology, Radiobiology, and the Biological Action of Magnetic Fields, Tomsk, p. 375.

Él'darov, A. L., and Kholodov, Yu. A. (1964). The effect of a constant magnetic field on motor activity, Zh. Obshch. Biol., 25:224.

Eliseev, V. V. (1964). Method of irradiating animals in experimental investigations of the action of radio-frequency electromagnetic waves, in· The Biological Action of Radio-Frequency Electromagnetic Waves, Moscow, p. 94.

Ellison M. A. (1955). The Sun and Its Influence, Routledge, London.

Ely, T., and Goldman, D. (1956). Heat exchange characteristics of animals exposed to 10-cm microwaves, Trans. IRE Med. Electronics, GM-4:38.

Ely, T., and Goldman, D. (1959). Heating characteristics of laboratory animals exposed to ten-centimetre microwaves, Nav. Med. Res. Inst., 15:77.

Encyclopedic Physics Dictionary (1965). Vol. 4, Izd. "Sovetskaya éntsiklopediya," Moscow.

England, T. (1950). Dielectric properties of the human body for wavelengths in the 1-10 cm range, Nature, 166:480.

England, T., and Sharpless, N. (1949). Dielectric properties of the human body in the microwave region of the spectrum, Nature, 163:487.

Epstein, N., and Cook, H. (1951). The effects of microwaves on the Rows N 1 fowl sarcoma virus, Brit. J. Cancer, 5:244.

Esay, A., et al. (1936). Temperaturmessungen angeschichteter biologischer Geweben bei Frequenzen von $2.7 \cdot 10^7$ Hz bis $1.2 \cdot 10^9$ Hz, Naturwissenschaften, 24:520.

Everdingen, W. Van (1938). Bestralingen met golven van Hertz, Ned. Tijdschr. Geneesk., 82:284.

Everdingen, W. Van (1940). Moleculare veranderingen tengevolge van bestraling met golven van Hertz met een frequentie van 1875 Megahertz, Ned. Tijdschr. Geneesk., 84:4370.

Everdingen, W. Van (1941). Moleculare veranderingen en strukturwjzigungen ten gevolge van bestralingen met golven van Hertz met een goltlengte van 10 centimeter, Ned. Tijdschr. Geneesk., 85·3094.

Everdingen, W. Van (1946a). Sur l'altération moléculaire et structurale par irradiation avec des ondes hertziennes de 16 et 10 centimètres (1875 et 3000 mHz), Rev. Belge Sci. Med., 5:272.

Everdingen, W. Van (1946b). Sur l'altération moléculaire et structurale par irradiation avec des ondes hertziennes de 16 et 10 centimètres (1875 et 3000 MHz): métabolisme hépatique et problème du cancer, Rev. Belge Sci. Med., 5:279.

Ewart, A. (1903). On the Physics and Physiology of Protoplasmic Streaming in Plants, Clarendon Press, Oxford.

Fessard, A., and Szabo, T. (1961). Mise en évidence d'un récepteur sensible a l'électricité dans la peau des mormyres, Compt. Rend., 253:1859.

Fleming, H. (1944). Effect of high-frequency fields on microorganisms, Electr. Eng., 1:18.

Fleming, J., et al. (1961). Microwave radiation in relation to biological systems and neural activity, in: Biological Effects of Microwave Radiation, Plenum Press, New York, p. 239.

Foerster, H. von (1962). Bio-logic, in: Biological Prototypes and Synthetic Systems, Plenum Press, New York.

Foner, B. (1963). Human erythrocyte agglutination in magnetic fields, Abstr. WE-4, Biophys. Soc., Seventh Annual Meeting.

Franke, V. A (1957). Measurement of electric and magnetic components of a high-frequency field in the immediate vicinity of radiation sources (in the induction zone) in the range 100 kHz-300 MHz, in· Proceedings of Jubilee Scientific Session of Institute of Labor Hygiene and Occupational Diseases of Academy of Medical Sciences of the USSR, Moscow, p. 71.

Franke, V. A. (1958). Measurement of electric and magnetic components of a high-frequency field in the frequency range 100 kHz-3 MHz and the design of equipment, in: Protection from the Action of Electromagnetic Fields and Electric Current in Industry, Leningrad, p. 64.

Franke, V. A., et al. (1962). Study of high-frequency components in electrocardiograms by power spectrum analysis, Circulation Res., 10:870.

Frenkel', G. L. (1930-1940). The UHF Electric Field in Biology and Experimental Medicine, Nos. 1-4, Medgiz, Moscow.

Frey, A. (1961). Auditory system response to radio-frequency energy, Aerospace Med., 32:1140.

Frey, A. (1962). Human auditory system response to modulated electromagnetic energy, J. Appl. Physiol., 17:689.

Frey, A. (1963a). Some effects on human subjects of ultrahigh-frequency radiation, Am. J. Med. Electronics, 2:28.

Frey, A. (1963b). Human response to very low-frequency electromagnetic energy, Naval Res. Rev., 16:1.

Frey, A. (1965). Behavioural biophysics, Psychol. Bull., 63:322.

Fricke, H. (1925). A mathematical treatment of the electric conductivity and capacity of a disperse system, Phys. Rev., 26:678.

Friedman, H., et al. (1962). Direct current potentials in hypnoanalgesia, Gen. Psychol., 79:193.

Friedman, H., and Becker, R. (1963). Geomagnetic parameters and psychiatric hospital admissions, Nature, 200:626.

Fukalova, P. P. (1964a). A hygienic assessment of conditions of work with SW and USW sources on radio and television stations, in: The Biological Action of Radio-Frequency Electromagnetic Fields, Moscow, p. 158.

Fukalova, P. P. (1964b). The effect of short and ultrashort waves on the body temperature and survival of experimental animals, in: The Biological Action of Radio-Frequency Electromagnetic Fields, Moscow, p. 78.

Fukalova, P. P. (1964c). Sensitivity of the olfactory analyzer in persons exposed to continuous SW and USW, in: The Biological Action of Radio-Frequency Electromagnetic Fields, Moscow, p. 144.

Furedi, A., and Valentine, R. (1962). Factors involved in the orientation of microscopic particles in suspensions influenced by radio-frequency fields, Biochim. Biophys. Acta, 56:33.

Furedi, A., and Ohad, I. (1964). Effects of high-frequency electric fields on the living cell, Biochim. Biophys. Acta, 79·1.

Gapeev, P. I. (1957). The effect of a SHF field on the visual organ, Tr. VMOLA, 73:152.

Gel'fon, I. A. (1964). The effect of 10-cm waves of low intensity on the histamine content of the blood of animals, in· The Biological Action of Radio-Frequency Electromagnetic Fields, Moscow, p. 68.

Gel'fon, I. A., and Sadchikova, M. N. (1964). Protein fractions and histamine in the blood after exposure to radiowaves of various ranges, in: The Biological Action of Radio-Frequency Electromagnetic Fields, Moscow, p. 133.

Gengerelli, J., et al. (1961). Magnetic fields accompanying transmission of nerve impulses in the frog sciatic, J. Psychol., 52·2.

Gengerelli, J., et al. (1964). Further observations on the magnetic field accompanying nerve transmission and tetanus, J. Psychol., 54:201.

Gerard, R. (1966). General results, in: The Concept of Information and Biological Systems, Mir, Moscow.

Gerencser, V., et al. (1962). Inhibition of bacterial growth by magnetic fields, Nature, 196:539.

Gerencser, V., et al. (1964). Inhibtion of bacterial growth in fields of high paramagnetic strength, in: Biological Effects of Magnetic Fields, Vol. 1, Plenum Press, New York, p. 229.

Ginetsinskii, A. G., and Lebedinskii, A. V. (1956). Course of Normal Physiology, Medgiz, Moscow.

Ginzburg, D. A., and Sadchikova, M. N. (1964). Changes in the electroencephalogram due to chronic exposure to radiowaves, in: The Biological Action of Radio-Frequency Electromagnetic Fields, Moscow, p. 126.

Glebov, N. A., et al. (1965). The effect of water treated with a magnetic field ("ÉLANG") on laboratory animals, in: Questions of Hematology, Radiobiololgy, and the Biological Action of Magnetic Fields, Tomsk, p. 390.

Glezer, V. D., and Tsukerman, I. I. (1961). Information Theory and Vision, Izd. Akad. Nauk SSSR, Moscow.

Goldsmith, T. (1966). Biomedical technique, Zarubezhnaya radioélektronika, No. 8:38.

Gordon, D. (1948). Sensitivity of the homing pigeon to the magnetic field of the earth, Science, 108:710.

Gordon, Z. V. (1964a). Results of an all-round investigation of the biological action of electromagnetic waves of radio frequencies and the prospects of future research, in: The Biological Action of Radio-Frequency Electromagnetic Fields, Moscow, p. 3

Gordon, Z. V. (1964b). The effect of microwaves on blood pressure in experiments on animals, in: The Biological Action of Radio-Frequency Electromagnetic Waves, Moscow, p. 57.

Gordon, Z. V. (1966). Questions of Occupational Hygiene and the Biological Action of Superhigh-Frequency Electromagnetic Fields, Meditsina, Moscow.

Gordon, Z. V., and Eliseev, V. V. (1964). Means of protection from SHF radiation and their effectiveness, in: The Biological Action of Radio-Frequency Electromagnetic Fields Moscow, p. 151.

Gordon, Z. V., and Lobanova, E. A. (1960). The temperature reaction of animals to exposure to SHF, in: The Fiological Action of Superhigh-Frequency Fields, Moscow, p. 59.

Gordon, Z. V., and Presman, A. S. (1965). Preventive and Protective Measures in Work with Centimeter-Wave Generators. Izd. BTI MRTP, Moscow.

Gordon, Z. V., et al. (1963). Materials on the biological action of microwaves of various ranges, Biol. i Medits. Élektronika, No. 6:72.

Gordy, W., et al. (1953). Microwave Spectroscopy, John Wiley and Son, New York.

Gorodetskaya, S. F. (1960). The effect of centimeter radiowaves on higher nervous activity and the hemopoietic and reproductive organs, Fiz. Zh. Akad. Nauk, 6:622.

Gorodetskaya, S. F. (1961). The effect of 8-centimeter radiowaves on the functional state of the adrenal cortex, Fiz. Zh. Akad. Nauk, 7:672.

Gorodetskaya, S. F. (1963). The effect of centimeter waves on the fecundity of female mice, Fiziol. Zh. Akad. Nauk. USSR, 9: 394.

Gorodetskaya, S. F. (1964a). The effect of an SHF electromagnetic field on the reproduction, peripheral blood composition, conditioned-reflex activity, and morphology of the internal organs of white mice, in: The Biological Action of Ultrasound and Superhigh-Frequency Electromagnetic Vibrations, Naukova Dumka, Kiev, p. 80.

Gorodetskaya, S. F. (1964b). The effect of an SHF field and convective heat on the estral cycle of mice, Fiziol. Zh. Akad. Nauk USSR, 10:494.

Grant, E. (1957). The dielectric method of estimating protein hydration, Phys. Med. Biol., 2:17.

Griffin, D. (1955). Bird navigation, in: Recent Studies in Avian Biology, University of Illinois Press, Urbana, p. 154.

Gross, L. (1962). Effect of magnetic fields on tumor immune responses in mice, Nature, 195:662.

Gross, L. (1963). The influence of magnetic fields on the production of antibody, in: Biomedical Sciences Instrumentation, Vol. I, Plenum Press, New York, p. 137.

Gross, L. (1964). Distortion of the bond angle in a magnetic field and its possible magnetobiological implication, in: Biological Effects of Magnetic Fields, Vol. 1, Plenum Press, New York, p. 74.

Grouch, G. (1948). Dielectric measurements at microwave frequencies, J. Chem. Phys., 16:364.

Gruzdev, A. D. (1965). The orientation of microscopic particles in electric fields, Biofizika, 10·1091.

Gulyaev, P. I. (1967). The electroauragram. The electric field of organisms as a new biological connection, in: Proceedings of Symposium on Physics and Biology, Moscow, p.19.

Gulyaev, P. I., Zabotin, V. P., and Shlippenbakh, N. Ya. (1967). The electric field in the air around excited tissues. The electroauragram, Paper read to the Leningrad Society of Naturalists, February 13.

Gunn, S., et al. (1961a). The effects of microwave radiation on the male endocrine system of the rat, Lab. Invest., 10:301.

Gunn, S., et al. (1961b). The effects of microwave radiation (24,000 Mc) on the male endocrine system of the rat, in: Biological Effects of Microwave Radiation, Vol. 1, Plenum Press, New York, p. 99.

Gvozdikova, Z. M., et al. (1964). The action of continuous electromagnetic fields on the central nervous system, in: The Biological Action of Radio-Frequency Electromagnetic Fields, Moscow, p. 20.

Hackel, E. (1946). Magnetic field effects on erythrocyte agglutination, Vox Sanguinis, 9:60.

Hackel, E., et al. (1961). Enhancement of human erythrocyte agglutination by constant magnetic fields, in: Second International Conference on High Magnetic Fields, Cambridge.

Hackel, E., et al. (1964). Agglutination of human erythrocytes, in Biological Effects of Magnetic Fields, Vol. 1, Plenum Press, New York, p. 218.

Haggis, G., et al. (1951). Estimation of protein hydration by electric measurements at microwave frequencies, Nature, 167:607.

Halberg, F. (1960). Temporal coordination of physiologic function, Cold Spring Harbor Symp. Quant. Biol. 25·289.

Handbook of Biological Data (1956). National Research Council, Washington.

Hansel, C. (1966). ESP — A Scientific Evaluation, Charles Scribner and Sons, New York.

Harary, J. (1962). Heart cells in vitro, Sci., Am., 206:141.

Hardy, A. (1949). Report on the 112th Annual Meeting of British Association for Advancement of Science, Science, 110:523.

Hartmuth, Z. (1954). "Die dielektrischen Eigenschaften biologischer Substanzen im Dezimeterwellen-Bereich," Z. Naturforsch., 96:257.

Hedrick, H. (1964). Inhibition of bacterial growth in homogeneous fields, in: Biological Effects of Magnetic Fields, Vol. 1, Plenum Press, New York, p. 220.

Heinmets, F., and Hershman, A. (1961). Consideration of the effects produced by superimposed electric and magnetic fields in biological systems and electrolytes, Phys. Med. Biol., 5:271.

Heller, J. (1959). Effect of high-frequency electromagnetic fields on microorganisms, Radio Electronics, 6:6.

Heller, J. (March 5, 1963). High-Frequency Treatment of Matter, U. S. Pat. 195,078, No. 3,095,359.

Heller, J., and Mickey, G. (1961). Non-thermal effects of radio frequency in biological systems, Digest Internat. Conf. Electronics, 21:2.

Heller, J., and Teixeira-Pinto, A. (1959). A new physical method of creating chromosomal aberration, Nature, 183:905.

Hendler, E., and Hardy, J. (1960). Infrared and microwave effects on skin temperature sensation, Trans. IRE Med. Electronics, ME-7:143.

Herrick, J. (1958). Pearl chain formation, in: Proc. Second Tri-Serv. Conf. on Biological Effects of Microwave Energy, Rome,

New York, p. 83.

Herrick, J., et al. (1950). Dielectric properties of tissues important in microwave diathermy, Federation Proc., 9:60.

Herron, T. (1965). Phase modulation of geomagnetic micropulsations, Nature, 207:699.

Hill, T. (1958). Some possible biological effects of an electric field acting on nucleic acids or proteins, J. Am. Chem. Soc., 8:2142.

Hirsch, F. (1956). The use of biological stimulants in estimating the dose of microwave energy, Trans. IRE, Med. Electronics, PGME-4:22.

Hoeft, L. (1965). Microwave heating: a study of the critical exposure variables for man and experimental animals, Aerospace Med., 36:621.

Holman, H. (1952). Telepathic communications, Proc. IRE, 40:995.

Horn, G. (1965). Le caratteristische elettriche passive dei sistemi biologici, Automaz. Automat., 9:5.

Howland, J., et al. (1961). Biomedical aspects of microwave irradiation on mammals, in: Biological Effects of Microwave Radiation, Vol. 1, Plenum Press, New York, p. 261.

Howland, J., and Michaelson, S. (1964). The effect of microwaves on the biologic response to ionizing radiation, Ind. Med. Surg., 33:500.

Imig, C., et al. (1948). Testicular degeneration as a result of microwave irradiation, Proc. Soc. Exptl. Biol. Med. 69:382.

Imyanitov, I. M., and Chubarina, E. V. (1961). The structure and origin of the electric field of the atmosphere, in: Investigations of Clouds, Precipitation, and Thunderstorm Electricity, Izd. Akad. Nauk SSSR, Moscow, p. 239.

Ingram, D. (1955). Spectroscopy at Radio and Microwave Frequencies, Butterworths Sci. Pubs., London.

Jackson, W. (1946). The representation of dielectric properties and the principles underlying their measurements at centimeter wavelengths, Trans. Faraday Soc., 42A:91.

Jackson, W. (1949). Dielectric behavior of methyl palmetate, Nature, 164:486.

Jaski, T. (1960). Radio waves and life, Radio Electronics, 31:43.

Kalant, H. (1959). Physiological hazards of microwave radiation: a survey of published literature, Can. Med. Assoc. J., 81:575.

Kalashnikov, A. G. (1956). Electricity, Gos. Izd. Tekhniko-Teor. Lit., Moscow.

Kamenskii, Yu. I. (1964). The effect of microwaves on the functional state of the nerve, Biofizika, 9·695.

Kamenskii, Yu. I. (1967). The effect of microwaves on the kinetics of nerve impulse parameters, Tr. Mosk. Obshchst. Usp'tatel. Prirod. (in press).

Karmilov, V. I. (1948). The history of the question of the biological and therapeutic effect of a magnetic field, in: The Biological and Therapeutic Effect of a Magnetic Field and Strictly Periodic Vibration, Perm, p.5.

Kay, C., and Schwan, H. (1957). Capacitive properties of body tissues, Circulation, 5:439.

Kelly, M. (1962). Electromagnetic effects on the nervous system, Research Report 63-27, University of California, Berkeley.

Kelly, M. (1963). Measurement of neuron activity by paramagnetic resonance, Research Report N PIBMPI-1071-62, Polytechnic Institute, Brooklyn, New York.

Keplinger, M. (1958). Review of the work conducted at the University of Michigan, in: Proc. Second Tri-Serv. Conf. on Biological Effects of Microwave Energy, Rome, New York, p. 215.

Kerova, N. I. (1964). The effect of an SHF electromagnetic field on polynuclease activity and nucleic acid content, in: The Biological Action of Ultrasound and Superhigh-Frequency Electromagnetic Vibrations, Naukova Dumka, Kiev, p. 108.

Khalifman, I. A. (1965). Bees, ants, and the Geiger counter, Nauka i zhizn, No. 5:72.

Kholodov, Yu. A. (1958). Formation of conditioned reflexes in fish to a magnetic field, in: Proceedings of Conference on the Physiology of Fish, Izd. Akad. Nauk SSSR, p. 82.

Kholodov, Yu. A. (1963a). The role of the main divisions of the brain of fish in the elaboration of electric defense conditioned reflexes to different stimuli, in: Nervous Mechanisms of Conditioned-Reflex Activity, Izd. Akad. Nauk Moscow, p. 287.

Kholodov, Yu. A. (1963b). Some features of the physiological action of electromagnetic fields from the results of conditioned-reflex and electroencephalographic techniques, in: Proceedings of Twentieth Conference on Problems of Higher Nervous Activity, Izd. Akad. Nauk SSSR, Moscow, p. 253.

Kholodov, Yu. A. (1965a). Progress in modern magnetobiology, in: Questions of Hematology, Radiobiology, and the Biological Action of Magnetic Fields, Tomsk, p. 309.

Kholodov, Yu. A. (1965b). The magnetic field as a stimulus, in: Bionics, Nauka, Moscow, p. 278.

Kholodov, Yu. A. (1966). The Effect of Electromagnetic and Magnetic Fields on the Central Nervous System, Nauka, Moscow.

Kholodov, Yu. A., and Zenina, I. N. (1964). The effect of caffeine on the EEG response to the action of a pulsed SHF field on the intact and isolated rabbit brain, in: The Biological Action of Radio-Frequency Electromagnetic Fields, Moscow, p. 33.

Khveledidze, M. A., et al (1965). The bioelectromagnetic field, in: Bionics, Nauka, Moscow, p. 305.

Kinosita, H. (1963). Electrical stimulation of Paramecium, J. Fac. Sci. Tokyo Univ., 4:137.

Kinosita, H. (1964). Electrical potentials and ciliary response in Opalina, J. Fac. Sci. Tokyo Univ., 7:1.

Kirchev, K., et al. (1962). Biochemical changes in the muscles and blood of white rats due to microwaves, in: Proceedings of the Fifth International Biochemical Congress, Section 14-28.

Kiryushkin, S. S., et al (1966). The relationship between developing eggs in a clutch and changes in them due to a weak magnetic field, in: Proceedings of Conference on the Effect of Magnetic Fields on Biological Objects, Moscow, p. 36.

Kitsovskaya, I. A. (1960). An investigation of the interrelationships between the main nervous processes in rats on exposure to SHF fields of various intensities, in: The Biological Effect of Superhigh Frequencies, Moscow, p. 75.

Kitsovskaya, I. A. (1964a). A comparison of the action of microwaves of different ranges on the nervous system of rats sensitive to sound stimulation, in: The Biological Action of Radio-Frequency Electromagnetic Fields, Moscow, p. 39.

Kitsovskaya, I. A. (1964b). The effect of centimeter waves of different intensities on the blood and hemopoietic organs of white rats, Gigiena Truda i Prof. Zabol., No. 6:14.

Klimkova-Deicheva, E., and Rot, B. (1963). The effect of radiation on the human encephalogram, Chekhoslov Med. Obozr., 9:254.

Klotz, I. (1962). Water, in: Horizons in Biochemistry, Academic Press, New York.

Knepton, J., and Beischer, D. (1964). The effects of magnetic fields

of up to 140,000 gauss on several organisms, Federation
Proc., 23(Pt. 1):523.

Knoepp, L., et al. (1962). The effect of low electrical frequencies
on various normal and malignant cells, Texas Rept. Biol.
Med., 20:623.

Knorre, K. G. (1960). Parameters of SHF fields for the assessment of working conditions and problems of their measurement, in: The Biological Action of Superhigh Frequencies,
Moscow, p. 11.

Knorre, A. G., and Lev, I. D. (1963). The Vegetative Nervous System, Medgiz, Moscow.

Kogan, A. B., and Tikhonova, N. A. (1965). The effect of a constant magnetic field on the movement of paramecia, Biofizika,
10:292.

Kogan, A. B., et al. (1965). The biological action of a constant magnetic field, in: Questions of Hematology, Radiobiology, and
the Biological Action of Magnetic Fields, Tomsk, p. 317.

Kogan, A. B., et al. (1966). The possible mechanism of action of
a constant magnetic field on the living cell, in: Proceedings
of Conference on the Effect of Magnetic Fields on Biological
Objects, Moscow, p. 37.

Kogan, I. M. (1966). Is telepathy possible?, Radiotekhnika, 21:8.

Konchalovskaya, N. M., et al (1964). The state of the cardiovascular system on exposure to radiowaves of various frequencies,
in: The Biological Action of Radio-Frequency Electromagnetic Fields, Moscow, p. 114.

Korteling, G., et al. (1964). Activity changes in alpha-amylase solution following their exposure to radio-frequency energy,
U. S. Army Med. Res. Lab., March 23, page 1.

Koryak, A. D. (1966). The effect of the geomagnetic field on the
vegetative growth of corn, in: Proceedings of Conference
on the Effect of Magnetic Fields on Biological Objects, Moscow, p. 43.

Krayukhin, B. V. (1948). Is electroinduction in the tissues of an
organism possible?, Collection Dedicated to the Memory of
A. V. Leontovich, Izd. Akad. Nauk SSSR, Kiev.

Krylov, A. V., and Tarakanova, G. A. (1960). Magnetotropism in
plants and its nature, Fiziol. Rast., 7:191.

Krylov, A. V., and Tarakanova, G. A. (1961). Magnetotropism,
Nauka i Zhizn', No. 7:85.

Kulakova, V. V. (1964). The effect of microwaves in the centimeter and decimeter ranges on the general and specialized forms of appetite in animals, in: The Biological Action of Radio-Frequency Electromagnetic Fields, Moscow, p. 70.

Kulin, E. T.(1965). Concentration and radio-frequency dependence of autoregulation of functions of unicellular organisms (paramecia), in: Papers on the Physicochemical Basis of Autoregulation in Cells, Moscow, p. 26.

Kulin, E. T., and Morozov, E. I. (1964). The effect of decimeter radio-emission on the phagocytic functions of unicellular organisms, Dokl. Akad. Nauk BSSR, 8:329.

Kulin, E. T., and Morozov, E. I. (1965). Some features of the effect of electromagnetic fields of the SHF range on the phagocytic function of paramecia, Vestn. Akad. Nauk BSSR, Ser. Biologich, Nauk, 4:91.

Laird, E. (1952). Dielectric properties of some solid proteins at wavelengths 1.7 m and 3.2 cm," Can. J. Phys., 30:663.

Laird, E., and Ferguson, K. (1949). Dielectric properties of some animal tissues at meter and centimeter wavelengths, Can. J. Res., Sec. A. 27:218.

Landau, L. D., and Livshits, E. M. (1957). Electrodynamics of Continuous Media, Gos. Izd. Tekhniko-Teor. Lit., Moscow.

Lang, O., and Koller, G. (1956). Schutzmasnahmen bei Hochfrequenzanlagen in Arbeits- Räumen," Zenth. Arbeitsmed. Arbeitsschutz, 6:13.

Lantsman, M. N. (1965). The effect of an alternating magnetic field on the phagocytic function of the reticulo-endothelial system in experiments, in: Questions of Hematology, Radiobiology, and the Biological Action of Magnetic Fields, Tomsk, p. 360.

Leary, F. (1959). Researching microwave health hazards, Electronics, 32:49.

Lebedeva, S. A. (1963). Measurement of a high-intensity electric field of industrial frequency and the currents flowing through a man to earth, in: Papers on Labor Hygiene and the Biological Action of Radio-Frequency Electromagnetic Fields, Moscow, p. 51.

Leites, F. L., and Skurikhina, L. A. (1961). The effect of microwaves on the hormonal activity of the adrenal cortex, Byul. Eksperim. Biol. i Med., No. 12:47.

Levengood, W. (1965). Factors influencing biomagnetic environment during the solar cycle, Nature, 205:465.

Levengood, W., and Shinkle, M. (1962). Solar flare effects on living organisms in magnetic fields, Nature, 195:967.

Levitina, N. A. (1961). Normal variation of the excitability of the neuromuscular system of rabbits, Fiziol. Zh. SSSR, 47:520.

Levitina, N. A. (1964). Effect of microwave irradiation of local regions of the body on the heart rate of the rabbit, Byul. Eksperim. Biol. i Med., No. 7:67.

Levitina, N. A. (1966a). An investigation of the nonthermal action of microwaves on the frog heart rate, Byul. Eksperim. Biol. i Med., No. 12:64.

Levitina, N. A. (1966b). Author's Abstract of Candidate's Dissertation: An Investigation of the Nonthermal Action of Microwaves on the Heart Rate, Moscow.

Libezin, P. (1936). Short and Ultrashort Waves in Biology and Therapy, Biomedgiz, Moscow.

Liboff, R. (1965). A biomagnetic hypothesis, Biophys. J., 5:845.

Lindauer, M., and Martin, H. (1968). Die Schweorientierung der Bienen unter dem Einfluss des Erdmagnetfeldes, Z. Vergleich. Physiol., 60:219.

Lissman, H. (1958). On the function and evolution of electric organs in fish, J. Exptl. Biol., 35:156.

Lissman, H., and Machin, K. (1958). The mechanism of object location in *Gymnarchus niloticus* and similar fish, J. Exptl. Biol., 35:451.

Lissman, H., and Machin, K. (1963). Electric receptors in a nonelectric fish, Nature, 199:88.

Livenson, A. R. (1962). Dosimetric methods in microwave therapy, Proceedings of Second All-Union Conference on the Use of Radioelectronics in Biology and Medicine, Moscow, p. 25.

Livenson, A. R. (1963). Dosimetric methods in centimeter and decimeter-wave therapy, Tr. Vses. Nauch.-Ussled. Inst. Med. Inst. i Oborudovaniya, 3:12.

Livenson, A. R. (1964a). Electric parameters of biological tissues in the microwave range, Meditsinskaya Promyshlennost', No. 7:10.

Livenson, A. R. (1964b). Questions of occupational hygiene relating to the operation of equipment for microwave therapy, Vopr. Kurortol., Fizioterapii i Lecheb. Fiz. Kul't., No. 5:450.

Livshits, N. N. (1957a). The role of the nervous system in the reactions of the organism to the action of UHF electromagnetic fields, Biofizika, 2:378.

Livshits, N. N. (1957b). Conditioned-reflex activity of dogs on local exposure of some zones of the cerebral cortex to a UHF field, Biofizika, 2:197.

Livshits, N. N. (1957c). Conditioned-reflex activity of dogs on exposure of the cerebellum region to a UHF field, Dokl. Akad. Nauk SSSR, 112:145.

Livshits, N. N. (1958). The role of the nervous system in the reactions of the organism to the action of a UHF electromagnetic field, Biofizika, 3:409.

Lobanova, E. A. (1959). Changes in the conditioned-reflex activity of animals (rats and rabbits) due to chronic exposure to centimeter waves, in: Papers on Labor Hygiene and the Biological Action of Radio-Frequency Electromagnetic Waves, Moscow, p. 46.

Lobanova, E. A. (1960). The survival and development of animals exposed to SHF fields of different intensity and duration, in: The Biological Action of Superhigh Frequencies, Moscow, p. 61.

Lobanova, E. A. (1964a). An investigation of the temperature reaction of animals to exposure to microwaves of various ranges, in: The Biological Action of Radio-Frequency Electromagnetic Fields, Moscow, p. 75.

Lobanova, E. A. (1964b). Changes in conditioned-reflex activity of animals due to exposure to microwaves of various frequency ranges, in: The Biological Action of Radio-Frequency Electromagnetic Fields, Moscow, p. 13.

Lobanova, E. A., and Gordon, Z. V. (1960). An investigation of the sense of smell in persons exposed to the action of SHF, in: The Biological Action of Superhigh Frequencies, Moscow, p. 52.

Lobanova, E. A., and Tolgskaya, M. S. (1960). Changes in the higher nervous activity and interneuronal connections in the cerebral cortex of animals due to the action of SHF, in: The Biological Action of Superhigh Frequencies Moscow, p. 69.

Lorente de No, R. (1947). A study of nerve physiology, Studies of Rockefeller Inst. Med. Res., 131:132.

Luk'yanova, S. N. (1960). The effect of a constant magnetic field on the bioelectric activity of the isolated nerve cord of the river crayfish, in: Proceedings of Conference on the Effect of Magnetic Fields on Biological Objects, Moscow, p. 45.

Luk'yanova, S. N. (1965). Changes in the electric activity of various brain formations of the rabbit due to the action of a constant magnetic field, in: Questions of Hematology, Radiobiology, and the Biological Action of Magnetic Fields, Tomsk, p. 368.

McAfee, R. (1961). Neurophysiological effect of 3-cm microwave radiation, Am. J. Physiol., 200:192.

McAfee, R. (1962). Physiological effects of thermode and microwave stimulation of peripheral nerves, Am. J. Physiol., 203:374.

McAfee, R., et al. (1961). Neurological effects of 3-cm microwave irradiation, in: Biological Effects of Microwave Radiation, Vol. 1, Plenum Press, New York, p. 251.

McCulloch, W. (1962). The imitation of one form of life by another — biomimesis, in: Biological Prototypes and Synthetic Systems, Plenum Press, New York.

McElroy, W., and Seliger, A. (1962). Biological luminescence, Sci. Am., 5:2.

McIlwain, H. (1959). Biochemistry and the Central Norvous System, Little, Brown and Co., Boston.

Magoun, H. W. (1958). The Waking Brain, Charles C. Thomas, Springfield, Illinois.

Malakhov, A. N., et al. (1963). A SHF electromagnetic field as a signal factor in a conditioned defense reflex of white mice, in: Proceedings of the Third Volga Conference of Physiologists, Biochemists, and Pharmacologists, Gorky, p. 310.

Malakhov, A. N., et al. (1965a). Biological indication of a SHF electromagnetic field, in: Bionics, Nauka, Moscow, p. 302.

Malakhov, A. N., et al. (1965b). The electromagnetic hypothesis of biological communication, in: Bionics, Nauka, Moscow, p. 297.

Maletto, S., et al. (1965a). Il valore degli acidi grassi a catena breve e con numero pari di atomi di carbonio (da C^4 a C^{14}) presenti nella quota lipidica del latte secreto da bovine sottoposte, a livello mammaria, all'azione de un campo magnetico rotante in bassa frequenza, Atti Fac. Med. Veter. Torino Vol. 19.

Maletto, S., et al. (1965). Gli acidi grassi della quota lipidica del latte vaccino secreto dalla ghiandola mammaria sottoposta all'azione di un campo manetico rotante in bassa frequenza, Atti Fac. Med. Veter. Torino, Vol. 19.

Maletto, S., et al. (1965c). La quantita de acidi grassi saturi e di acidi grassi insaturi e le variazioni dell nidice AGS/AGI nella quota lipidica del latte vaccino secreto dalla ghiandola mammaria cimentata da un campo magnetico rotante in bassa frequenza, Atti Fac. Med. Veter. Torino, Vol. 19.

Maletto, S., ánd Valfre, F. (1966). La nuovo ecologia i osservazioni e rilievi sperimentali dell'ordine zootecnico, Report of the Fourth International Biometeorological Congress.

Maletto, S., et al. (1966a). Latte alimentare prodotto dalla ghiandola mammaria sottoposta all'azione di un campo magnetico di B. F., Minerva Pediat. 18:784.

Maletto, S., et al. (1966b). Alcune costanti fisiche e chimiche del latte alimentare prodotto dalla ghiandola mammaria sottoposta all'azione di un campo magnetico di B. F., Minerva Pediat. 18:788.

Mancharskii, S. (1964). New ways of acting on the human sense organs, Zarubezhnaya Radioélektronika, No. 7:52.

Manteifel', B. P., et al. (1965). Orientation and navigation in the animal world, in: Bionics, Nauka, Moscow, p. 245.

Manual of Geophysics (1965). Nauka, Moscow.

Marha, K. (1963). Nektera experimentalni pozorovani ucinku vysokofrekvencniho electromgnetickeho pole in vivo a in vitro, Procavni Lekar, 15:238.

Marikovskii, P. (1965). Ant language, Nauka i Zhizn', No. 6:55.

Masoero, P., et al. (1965). Primi risultati circa 1, influenza dei campi elettrostatici e dell'acqua "attivata" sull'accrescemento ponderale, Minerva Pediat. 17:1133.

Masoero, P., et al. (1966). Hipoteza o novoj ekologiji, Stocarstvo, 20:29.

Maw, M. (1961). Suppression of oviposition rate of *Scambus bidianae* (Hymenoptera, Ichneumonidae) in fluctuating electric fields, Can. J. Entomol. 93:602.

Mekshenkov, M. I. (1965a). An investigation of the structure and conformation of ribonucleic acids by the method of birefringence in a magnetic field, Biofizika, 10:747.

Mekshenkov, M. I. (1965b). Magnetic double refraction in a DNA solution, in: Molecular Biophysics, Nauka, Moscow, p. 155.

Mericle, R., et al. (1964). Plant growth response, in: Biological Effects of Magnetic Fields, Vol. 1, Plenum Press, New York, p. 183.

Merli, G., et al. (1963). Azione del calore e delle onde corte sulla cellula neoplastica, Arch. Ital. Patol. Clin. Tumori, 6:57.

312 REFERENCES

Mermagen, H. (1961). Phantom experiments with microwaves at the University of Rochester, in: Biological Effects of Microwave Radiation, Vol. 1, Plenum Press, New York, p. 143.

Merola, L., and Kinosita, J. (1961). Changes in the ascorbic acid content in lenses of rabbit eyes exposed to microwave radiation, in: Biological Effects of Microwave Radiation, Vol. 1, Plenum Press, New York, p. 285.

Michaelson, S., et al. (1958). The biological effects of microwave irradiation in the dog, Proc. Second Tri-Serv. Conf. on Biological Effects of Microwave Energy, Rome, New York, p. 175.

Michaelson, S., et al. (1961a). The biological effects of microwave irradiation, Digest Internat. Conf. Med. Electronics, 26:4.

Michaelson, S., et al. (1961b). Tolerance of dogs to microwave exposure under various conditions, Ind. Med. Surg., 30:298.

Michaelson, S., et al. (1961c). Physiological aspects of microwave irradiation of mammals, Am. J. Physiol., 201:351.

Michaelson, S., et al. (1963). The influence of microwaves on ionizing radiation exposure, Aerospace Med., 34:111.

Michaelson, S., et al. (1964). The hematologic effects of microwave exposure, Aerospace Med., 35:824.

Mickey, G. (1963). Electromagnetism and its effect on the organism, N. Y. State J. Med., 63:1935.

Minecki, L., and Romanik, A. (1963a). Zmiany czyunosci odruchowo warunkowej szczorow pod wplywen dzialania mikrofal (Plasmo "S"). I. Jednrazowe dzialanie mikrofal, Med. Pracy, 4:255.

Minecki, L., and Romanik, A. (1963b). Zmiany czyunosci odruchowo warunlowej szczurow pod wplywen dzialania mikrofal (Plasmo "S"). II. Przewlekle dzialanie mikrofal, Med. Pracy, 14:361.

Minenko, V. I. et al. (1962). Magnetic Treatment of Water, Khar'kovskoe Knizhnoe Izd.

Mirutenko, V. I. (1964). The thermal effect of a SHF electromagnetic field on animals and some questions of SHF-field dosimetry, in: The Biological Action of Ultrasound and Super-high-Frequency Electromagnetic Vibrations, Naukova Dumka, Kiev, p. 62.

Mittelman, E. (1961). Relationship between heat sensation and high-frequency power absorption by humans, Digest Internat. Conf. Med. Electronics, 26:3.

Mogendovich, M. R. (1965). Magnetic field and physiological functions, in: Questions of Hematology, Radiobiology, and the Biological Action of Magnetic Fields, Tomsk, p. 314.

Mogendovich, M. R., and Sherstneva, O. S. (1947). Gravitational effects of blood in a magnetic field, Byull. Eksperim. Biol. i Med., No. 12:459.

Mogendovich, M. R., and Sherstneva, O. S. (1948a). Erythrocyte sedimentation rate in a magnetic field, in: The Biological and Therapeutic Action of a Magnetic Field and Strictly Periodic Vibration, Perm, p. 61.

Mogendovich, M. R., and Sherstneva, O. S. (1948b). Gravitational effect of blood in a magnetic field, in: The Biological and Therapeutic Action of a Magnetic Field and Strictly Periodic Vibration, Perm, p. 73.

Mogendovich, M. R., and Skachedub, R. G. (1957). The effect of physical factors on the human visual apparatus, Tr. Permsk. Med. Inst., No. 26:11.

Mogendovich, M. R., and Tishankin, V. F. (1948a). Mechanism of the effect of a magnetic field on the ESR, in: The Biological and Therapeutic Action of a Magnetic Field and Strictly Periodic Vibration, Perm, p. 79.

Mogendovich, M. R., and Tishankin, V. F. (1948b). Mechanism of the effect of a magnetic field on the ESR, Byull. Eksperim. Biol. i Med., No. 6:417.

Montani, A. (1944). A generator of damped microwaves, Electronics, 17:144.

Moos, W. (1964). A preliminary report on the effects of electric fields on mice, Aerospace Med., 35:374.

Moressi, W. (1964). Mortality patterns of mouse sarcoma 180 cells resulting from direct heating and chronic microwave irradiation, Exptl. Cell Res., 33:240.

Moskalenko, Yu. E. (1958). The use of SHF for biological investigations, Biofizika, 3:619.

Moskalenko, Yu. E. (1960). The clinical and biological use of SHF electromagnetic fields, in: Electronics in Medicine, Gosenergoizdat, Moscow, p. 207.

Moskalyuk, A. I. (1957). Effect of a SHF field on oxidation reduction processes in some rabbit tissues, Tr. VMOLA, 73:133.

Mulay, L. (1964). Basic concepts related to magnetic fields and magnetic susceptibility, in: Biological Effects of Magnetic Fields, Vol. 1, Plenum Press, New York, p. 33.

Mulay, L., and Mulay, I. (1961). Effect of a magnetic field on sarcoma 37 ascites tumor cells, Nature, 190:1019.

Mulay, L., and Mulay, I. (1964). Effects of *Drosophila*
melanogaster and S-37 tumor cells, in: Biological Effects
of Magnetic Fields, Vol. 1. Plenum Press, New York, p. 146.

Mumford, W. (1961). Some technical aspects of microwave radiation hazards, Proc. IRE, 49:427.

Murayama, M. (1965). Orientation of sickled erythrocytes in a magnetic field, Nature, 206:420.

Murphy, J. (1942). The influence of magnetic fields on seed germination, Am. J. Botany Suppl., 29:15.

Murr, L. (1965). Biophysics of plant growth in an electrostatic
field, Nature 206:467.

Naumov, N. P., and Il'ichev V. D. (1965). A few notes on the orientation problem, in: Bionics, Nauka, Moscow, p. 378.

Neiman, M. S. (1964). Some fundamental questions of microminia-
turization. Part I, Radiotekhnika, No. 1:3.

Neiman, M. S. (1965a). Some fundamental questions of micromi-
niaturization. Part II, Radiotekhnika, No. 1:3.

Neiman, M. S. (1965b). Some fundamental questions of micromi-
niaturization. Part III, Radiotekhnika, No. 6:1.

Neprimerov, N. N., et al. (1966). The mechanism of the biological
action of magnetic fields, in: Proceedings of Conference on
the Effect of Magnetic Fields on Biological Objects, Moscow,
p. 48.

Neurath, P. (1964). Simple theoretical models for magnetic interactions with biological units, in: Biological Effects of Magnetic Fields, Vol. 1, Plenum Press, New York, p. 25.

Neville, J. (1955). An experimental study of magnetic factors possibly concerned with bird navigation, Dissertation Abstr.,
15:1855.

New biological effects of R-F (1959). Electronics, 32:38.

Nikogosyan, S. V. (1960). Effect of SHF on cholinesterase activity
in blood serum and organs of animals, in: The Biological
Action of Superhigh Frequencies, Moscow, p. 81.

Nikogosyan, S. V. (1964a). Effect of 10-cm waves on amount of
protein fractions in animal blood serum, Gigiena Truda i
Prof. Zabolevaniya, No. 9:56.

Nikogosyan, S. V. (1964b). An investigation of cholinesterase activity in the blood serum of animals chronically exposed to
microwaves, in: The Biological Action of Radio-Frequency

Electromagnetic Fields Moscow, p. 43.

Nikogosyan, S. V. (1964c). Effect of 10-cm waves on nucleic acid content of animal organs, in: The Biological Action of Radio-Frequency Electromagnetic Fields, Moscow, p. 66.

Nikonova, K. V. (1963). Author's Abstract of Candidate's Dissertation: Materials for a Hygienic Evaluation of High-Frequency Electromagnetic Fields (Medium- and Long-Wave Frequency Ranges), Moscow.

Nikonova, K. V. (1964a). Effect of a high-frequency electromagnetic field on the blood pressure and body temperature of experimental animals, in: The Biological Action of Radio-Frequency Electromagnetic Fields, Moscow, p. 61.

Nikonova, K. V. (1964b). Effect of high-frequency electromagnetic fields on functions of the nervous system, in: The Biological Action of Radio-Frequency Electromagnetic Fields, Moscow, p. 49.

Novak, J. (1963). Vliv impulsniho elektromagnetickeho pole na lidsky organismus, Casopis Lekaru Ceskych, 102:496.

Novitskii, Yu. I. (1966a). The action of a constant magnetic field on gels of substances of plant origin, in: Proceedings of Conference on the Effect of Magnetic Fields on Biological Objects Moscow, p. 50.

Novitskii, Yu. I. (1966b). Effect of a magnetic field on the dry seeds of some cereals, in: Proceedings of Conference on the Effect of Magnetic Fields on Biological Objects, Moscow, p. 52.

Novitskii, Yu. I., et al. (1965). A further study of the effect of a constant magnetic field on plants, in: Questions of Hematology, Radiobiology, and the Biological Action of Magnetic Fields, Tomsk, p. 329.

Novitskii, Yu. I., et al. (1966). Effect of a weak magnetic field on the movement of chloroplasts in Elodea, in: Proceedings of Conference on the Effect of Magnetic Fields on Biological Objects, Moscow, p. 53.

Nyrop, J. (1946). A specific effect of high-frequency electric currents on biological objects, Nature, 157:51.

Obrosov, A. N. (1963). A pulsed UHF field — a new therapeutic factor, in: Proceedings of First Republican Conference of Physiotherapists and Health-Resort Specialists of the Ukrainian SSR, Kiev, p. 238.

Obrosov, A., and Jasnogorodski, V. (Yasnogorodskii, V.) (1961). A new method of physical therapy — pulsed electric field of ultrahigh frequency, Digest Internat. Conf. Med. Electronics, p. 156.

Obrosov, A. N., et al. (1963). Effect of microwaves on cardiovascular system of a practically healthy man, Vopr. Kurortol., Fizioterapii i Lecheb. Fiz. Kul't., No. 3:223.

Odintsov, Yu. N. (1965). Effect of an alternating magnetic field on some immunological indices in experimental listiosis, in: Questions of Hematology, Radiobiology, and the Biological Action of Magnetic Fields, Tomsk, p. 382.

Orlova, A. A. (1960a). State of cardiovascular system on exposure to SHF and HF fields, in: Physical Factors of the External Environment Moscow, p. 171.

Orlova, A. A. (1960b). Clinical aspects of changes in internal organs due to exposure to SHF, in: The Biological Action of Superhigh Frequencies, Moscow, p. 36.

Osipov, Yu. A. (1965). Labor Hygiene and the Effect of Radio-Frequency Electromagnetic Fields on Workers, Medgiz, Moscow.

Paff, G., et al. (1962). The effects of microwave irradiation on the embryonic chick heart as revealed by electrocardiographic studies, Anat. Record,142:264.

Paff, G., et al. (1963). The embryonic heart subjected to radar, Anat. Proc, 147:379.

Palmer, J. (1962). Doctoral Dissertation: Effect of Weak Magnetic Fields on Spatial Orientation and Locomotor Responses of *Volvox aureus* , Northwestern Univ.

Palmer, J. (1963). Organismic spatial orientation in very weak magnetic fields, Nature, 198:1061.

Parducz, B. (1954). Reizphysiologische Untersuchungen an Zilliaten. I. Über das Aktions-System von Paramecium, Acta Microbiol. Acad. Sci. Hung., 1:175.

Pervushin, V. Yu., and Triumfov, A. V. (1957). Morphological changes in some organs of rabbits subjected to the action of a SHF field, Tr. VMOLA, 73:141.

Petrov, F. P. (1935). Effect of an electromagnetic field on isolated organs, in: Physicochemical Bases of Higher Nervous Activity, Leningrad, p. 97.

Petrov, F. P. (1952). Effect of a low-frequency electromagnetic field on higher nervous activity, Tr. Inst. Fiziol. Akad. Nauk SSSR, 1:369.

Petrovskii, A. A. (1926). Telepsychic phenomena and brain radia-
     tions, Telefoniya i Telegrafiya Bez Provodov, No. 34:61.
Piccardi, G. (1962). The Chemical Basis of Medical Climatology,
     Charles C. Thomas, Springfield, Illinois.
Piccardi, G. (1965a). Fluctuation phenomena — a subject for the
     world university, Third Plenary Meeting of World Academy
     of Art and Science, Rome.
Piccardi, G. (1965b). The universe everywhere, Nauka i Zhizn', No.
     8:65.
Picket, J., and Schrank, A. (1965). Responses of coleoptiles to
     magnetic and electric fields, Texas J. Sci., 17:245.
Pittman, U. (1962). Growth reaction and magnetotropism in roots
     of winter wheat, Can. J. Plant Sci., 42:430.
Pittman, U. (1963). Effects of magnetism on seedling growth of
     cereal plants, in: Biomedical Sciences Instrumentation, Vol.
     1, Plenum Press, New York, p. 117.
Plaksin, I. N., et al. (1966). Effect of frequency of electric field
     on optical properties of water, Dokl. Akad. Nauk SSSR, 168:1.
Plekhanov, G. F. (1965). Some materials on the reception of inform-
     ation by living systems, in: Bionics, Nauka, Moscow, p. 237.
Plekhanov, G. F., and Vedyushkina, V. V. (1966). Elaboration of a
     vascular conditioned reflex in man to change in strength of
     a high-frequency electromagnetic field, Vysshei Nervnoi
     Deyatel'nosti im. I. P. Pavlov, 16:34.
Poddubyi, A. G. (1965). Some results of long-range observations
     of behavior of migrating fish, in: Bionics, Nauka, Moscow,
     p. 255.
Polukhanov, M. P. (1965). The Propagation of Radio Waves, Svyaz',
     Moscow.
Povzhitkov, V. A., et al. (1961). Effect of a pulsed SHF electro-
     magnetic field on conception and course of gestation in white
     mice, Byull. Eksperim. Biol. i Med., No. 5:103.
Pozolotin, A. A. (1965). Mechanism of the different action of a
     magnetic field on irradiated pea seeds and seedlings, in:
     Questions of Hematology, Radiobiology, and the Biological
     Action of Magnetic Fields, Tomsk, p. 332.
Pozolotin, A. A., and Gatiyatullina, E. Z. (1966). The combined ef-
     fect of gamma-irradiation and a magnetic field on pea seeds,
     in: Proceedings of Conference on the Effect of Magnetic
     Fields on Biological Objects, Moscow, p. 58.

Prausnitz, S., et al. (1961). Longevity and cellular studies with microwaves, in: Biological Effects of Microwave Radiation, Vol. 1, Plenum Press, New York, p. 135.

Prausnitz, A., and Susskind, C. (1962). Effect of chronic microwave irradiation on mice, Trans. IRE, Med. Electronics, PGME-4: 104.

Presman, A. S. (1954a). Centimeter Waves, Gosenergoizdat, Moscow.

Presman, A. S. (1954b). An instrument for measuring the intensity of irradiation by 10-centimeter waves in industrial conditions, Annotations of Scientific Works of the Academy of Medical Sciences of the USSR, Moscow, p. 479.

Presman, A. S. (1956a). Physical bases of the biological action of centimeter waves, Usp. Sovr. Biolog., 41:40.

Presman, A. S. (1956b). An electromagnetic field as a hygienic factor, Gigiena i Sanit., No. 9:32.

Presman, A. S. (1957a). The hygienic evaluation of high-frequency electromagnetic fields, Proceedings of the Jubilee Scientific Session of the Institute of Labor Hygiene and Occupational Diseases, Moscow, p. 72.

Presman, A. S. (1957b). Methods of evaluating the effective energy of electromagnetic fields in industrial conditions, Gigiena i sanit., No. 1:29.

Presman, A. S. (1957c). Change in the human body and skin temperature due to irradiation with low-intensity centimeter waves, Byull. Eksperim. Biol. i Med., No. 2:51.

Presman, A. S. (1958a). Experimental methods of irradiating animals with centimeter waves, Biofizika, 3:354.

Presman, A. S. (1958b). Methods of protection against radio-frequency electromagnetic fields in industrial conditions, Gigiena i Sanit., No. 1:21.

Presman, A. S. (1960a). The hygienic evaluation of high-frequency electromagnetic fields, in: Physical Factors of the Environment, Moscow, p. 142.

Presman, A. S. (1960b). An experimental apparatus for measured irradiation of rabbits with microwaves in the 10-centimeter range, Novosti Med. Tekhniki, No. 4:51.

Presman, A. S. (1960c). The use of microwaves in physiotherapy and biological investigations, in: Electronics in Medicine, Gosenergoizdat, Moscow, p. 219.

Presman, A. S. (1961). An experimental apparatus for irradiating protein solutions with microwaves, Biofizika, 6:370.

Presman, A. S. (1962a). Methods of measured irradiation with microwaves in biological experiments, in: Proceedings of Second All-Union Conference on Use of Radioelectronics in Biology and Medicine, Moscow, p. 23.

Presman, A. S. (1962b). Mechanism of the nonthermal action of microwaves, in: Proceedings of the Second All-Union Conference on the Use of Radioelectronics in Biology and Medicine, Moscow, p. 21.

Presman, A. S. (1963a). A method of determining the excitation thresholds of the neuromuscular apparatus of animals, Biol. i Med. Elektronika, No. 5:56.

Presman, A. S. (1963b). The excitability of paramecia on stimulation with dc and ac pulses, Biofizika, 8:138.

Presman, A. S. (1963c). The effect of microwaves on paramecia, Biofizika, 8:258.

Presman, A. S. (1963d). A method of comparative irradiation of protein solutions with microwaves and infrared rays, Biol. i Med. Elektronika, No. 6:76.

Presman, A. S. (1963e). Mechanism of the biological action of microwaves, Usp. Sovr. Biol., 56:161.

Presman, A. S. (1964a). Investigations of the biological action of microwaves. Part I, Zarubezhnaya Radioelektronika, No. 3:63.

Presman, A. S. (1964b). Investigations of the biological action of microwaves. Part II, Zarubezhnaya radioelektronika, No. 4:67.

Presman, A. S. (1964c). The role of electromagnetic fields in vital processes, Biofizika, 9:131.

Presman, A. S. (1965a). The effect of microwaves on living organisms and biological structures, Usp. Fiz. Nauk, 86:263.

Presman, A. S. (1965b). The electromagnetic field and life, Nauka i Zhizn', No. 5:82.

Presman, A. S. (1966a). Electromagnetic fields in neurocybernetics, in: Proceedings of Symposium on Problems of Neurocybernetics, Moscow, p. 41.

Presman, A. S. (1966b). Some general methodological questions of bioelectromagnetic investigations, in: Proceedings of Conference on the Effect of Magnetic Fields on Biological Objects Moscow, p. 59.

Presman, A. S. (1967a). Electromagnetic fields and regulation processes in biology, in: Questions of Bionics, Nauka, Moscow, p. 341.

Presman, A. S. (1967b). The role of electromagnetic fields in evolution and the vital activity of organisms, Byull. Mosk. Obshchest. Isp'tatel. Prirod., 52:149.

Presman, A. S. (1967c). The interaction of physics and biology in the investigation of the biological effect of electromagnetic fields, in: Proceedings of Symposium on Physics and Biology, Moscow, p. 13.

Presman, A. S., and Kamenskii, Yu. I. (1961). Experimental apparatuses for investigation of the excitability of a nerve-muscle preparation during microwave irradiation, Biofizika, 6:231.

Presman, A. S., and Levitina, N. A. (1962a). Nonthermal action of microwaves on the heart rate of animals. I. Action of continuous microwaves, Byull. Eksperim. Biol. i Med., No. 1:41.

Presman, A. S., and Levitina, N. A. (1962b). Nonthermal action of microwaves on the heart rate of animals. II. Action of pulsed microwaves, Byull. Eksperim. Biol. i Med., No. 2:39.

Presman, A. S., and Levitina, N. A. (1962c). The effect of nonthermal microwave irradiation on the resistance of animals to gamma irradiation, Radiobiologiya, 2:170.

Presman, A. S., and Rappeport, S. M. (1964a). New data on the existence of an excitable system in paramecia. I. Reactions of paramecia to direct-current pulses, Nauchn. Dokl. Vysshei Shkoly, Ser. Biologichesk. Nauki, No. 1·48.

Presman, A. S., and Rappeport, S. M. (1964b). New data on the existence of an excitable system in paramecia. II. Reactions of paramecia to ac pulses, Nauchn. Dokl. Vysshei Shkoly, Ser. Biologichesk. Nauki, No. 3:44.

Presman, A. S., and Rappeport, S. M. (1965). Effect of microwaves on the excitable system of paramecia, Byull. Eksperim. Biol. i Med., No. 4:48.

Presman, A. S., et al. (1961). The biological action of microwaves, Usp. Sovr. Biol., 51:84.

Proceedings of First All-Union Conference on the Use of Ultrashort Waves in Medicine (1940). Moscow.

Pudovkin, A. I. (1964). Effect of a low-frequency electromagnetic field on frog nerve-muscle preparation, in: The Nervous System, Leningrad.

Radiation Cataracts (1959). Medgiz, Moscow.

Rae, J., et al. (1949). A comparative study of the temperature produced by microwave and short diathermy, Arch. Phys. Med., 30:199.

Rait, R. (1964). The command language which insects obey is of a physicochemical nature, Nauk i Zhizn', No. 5:148.

Rajewsky, V., and Schwan, H. (1948). Die Dielektrizitäts-Konstante und Leitfähigkeit des Blutes bei ultrahohen Frequenzen, Naturwissenschaften 10:315.

Ramo, S., and Whinnery J. (1944). Fields and Waves in Modern Radio, Wiley, New York.

Reiter, R. (1960). Meteorobiologie und Elektrizität der Atmosphäre, Akad. Verlagsgesellschaft, Leibzig.

Reno, V., and Nutini, L. (1963). Effect of magnetic fields on tissue respiration, Nature, 198:2040.

Reynolds, M. (1961). Development of a garment for protection of personnel working in high-power R-F environments, in: Biological Effects of Microwave Radiation, Vol. 1, Plenum Press, New York, p. 71.

Rhine, J., and Pratt, J. (1957). Parapsychology (Frontier Science of the Mind), Charles C. Thomas, Springfield, Illinois.

Richardson, A., et al. (1952). The role of energy, pupillary diameter, and alloxan diabetes in the production of ocular damage by microwave irradiation, Am. J. Ophthalmol., 35:993.

Riper, W. Van, and Kalmbach, E. (1952). Homing not hindered by wing magnets, Science, 115:577.

Riviere, M., et al. (1965a). Effets des champs électromagnétiques sur un lymphosarcome lymphoblastique transportable du rat, Semaine Hop. Inform., 11:6.

Riviere, M., et al. (1965b). Action des champs électromagnétiques sur les greffes de la tumeur T8 chez le rat, Semaine Hop. Inform., 11:3.

Riviere, M., et al. (1965c). Phénomènes de regression observés sur les greffes d'un lymphosarcome chez des souris exposées a des champs électromagnétiques, Compt. Rend., 260: 2639.

Riviere, M., et al. (1965d). Effet des champs électromagnétiques sur un lymphosarcome lymphoblastique transplantable du rat, Compt. Rend. 260:2099.

Roberts, J. (1959). Nuclear Magnetic Resonance, McGraw-Hill, New York.

Roberts, J., and Cook, H. (1952). Microwaves in medicine and research, Brit. J. Appl. Phys., 3:33.

Rodicheva, E. K., et al. (1965). The effect of constant electric and alternating electromagnetic fields on the biosynthesis of Chlorella in continuous cultivation, in: Questions of Hematology, Radiobiology, and the Biological Action of Magnetic Fields, Tomsk, p. 319.

Roppel, R. (1963). A study of near infrared emission from the mammalian cerebral cortex, Rept. DDC 417125, Defense Docum. Center, Alexandria, Virginia.

Ruderfer, M. (1952). Telepathy and the quantum, Proc. IRE, 40: 1735.

Ryzhov, A. I., and Garganeev, G. P. (1965). Pathomorphological changes in the gastrointestinal tract of animals due to a constant magnetic field, in: Questions of Hematology, Radiobiology, and the Biological Action of Magnetic Fields, Tomsk, p. 349.

Ryzl, M. (1962). Training of ESP by means of hypnosis, J. Soc. Psych. Res., No. 711.

Sacchitelli, G., and Sacchitelli, F. (1956). L'azione delle microonde radar sulla plasmalipasi e sull'amilasi serica, Folia Med., 39:1037.

Sacchitelli, G., and Sacchitelli, F. (1958). Sul compartmento della glutationemia in seguito ad irradiazioni con microonde radar, Folia Med., 41:342.

Sadchikova, M. N. (1964). Clinical aspects of changes in nervous system due to radiowaves in various ranges, in: The Biological Action of Radio-Frequency Electromagnetic Fields, Moscow, p. 110.

Salei, A. I. (1964). The effect of the energy of an electromagnetic field of various frequencies on salivary gland secretion, in: Some Questions of Physiology and Biophysics, Voronezh, p. 50.

Salzberg, B. (1966). What is information theory? in: The Concept of Information and Biological Systems, Mir, Moscow, p. 13.

Samoilov, O. Ya. (1957). The Structure of Aqueous Solutions of Electrolytes and the Hydration of Ions, Izd. Akad. Nauk SSSR, Moscow.

Sarel, M., et al. (1961). Zur Wirkung der elektromagnetischen Zentimeterwellen auf das Nervensystem des Menschen (Radar), Z. Ces. Hyg., 7:897.

Satio, M., et al. (1961a). R-F-field-induced forces on microscopic particles, Digest. Internat. Conf. Med. Electronics, 21:3.

Satio, M., et al. (1961b). The time constants of pearl chain formation, in: Biological Effects of Microwave Radiation, Vol. 1, Plenum Press, New York, p. 85.

Savostin, P. V. (1928). Behavior of rotating plant protoplasm in a constant magnetic field, Izv. Tomsk. Gos. Univ., 79:207.

Savostin, P. V. (1937). Magnetophysiological effects in plants, Tr. Mosk. Doma Uchenykh, No. 1:111.

Sazonova, T. E. (1960). The stimulating effect of a low-frequency (50 Hz) electromagnetic field on a nerve-muscle preparation, Tr. Leningr. Obehchest. Isp'tatel. Prirod. 71:84.

Sazonova, T. E. (1964). Author's Abstract of Candidate's Dissertation: Functional Changes in Organism Due to Work in a High-Intensity Electric Field of Industrial Frequency, Leningrad.

Scelsi, B. (1957). Termogenesi da ultrasuoni e microonde (onde Radar) nei tessuti organici non viventi, Radioterap. Radiobiol. Fis. Med., 12:135.

Schastnaya, P. I. (1955). The effect of SHF fields on microorganisms, in: Collection of Scientific Works of Kharkov Medical Institute, Kharkov, p. 170.

Schastnaya, P. I. (1957). The effect of electromagnetic waves of superhigh frequency on microorganisms, Tr. Khar'kovsk. Med., Inst. 15:239.

Schastnaya, P. I. (1958). The effect of SHF radiowaves on the colon bacillus, Tr. Khar'kovsk. Med. Inst., 16:359.

Scheminzky, Fe., Scheminzky, Fr., and Bukatsch, F. (1941). Electro-Tropismus, Electro-taxis, Electronarcose und verwandte Erscheinungen, Tab. Biol., 19:2.

Schmitt, F. O. (1959). Molecular biology and the physical basis of life processes, Rev. Mod. Phys., 31:5

Schua, L. (1953). Die Fluchtreaktionen von Goldhamstern aus elektrischen Feldern, Naturwissenschaften, 40:514.

Schwan, H. (1948). Electrical properties of living tissues, Am. J. Med. Sci., 215:233.

Schwan, H. (1950). Resonance method for the determination of complex resistances and substances at decimeter waves, Ann. Phys., 6:253.

Schwan, H. (1953a). Electrical properties of blood at ultrahigh frequencies, Am. J. Phys. Med., 32:144.

Schwan, H. (1953b). Messung von elektrischen Materialkonstanten und Komplex-Widerständen von allen biologischen Substanzen, Z. Naturforsch., 8:3.

Schwan, H. (1954). Die elektrischen Eigenschaften von Muskelgewebe bei Niederfrequenz, Z. Naturforsch., 98:245.

Schwan, H. (1955). Electrical properties of body tissues and impedance plethismography, Trans. IRE, Med. Electronics, PGME-3:32.

Schwan, H. (1956). Electrical properties measured with alternating current: body tissues, in: Handbook of Biological Data, Nat. Res. Council, Washington, D. C.

Schwan, H. (1957). Electrical properties of tissue and cell suspensions, Advan. Biol. Med. Phys., 5:147.

Schwan, H. (1958). Molecular response characteristics to ultrahigh frequency fields, Proc. Second Tri-Serv. Conf. on Biol. Effects of Microwave Energy, Rome, New York, p. 33.

Schwan, H. (1959). Alternating current spectroscopy of biological substances, Proc. IRE, 47:1841.

Schwan, H., and Carstensen, E. (1957). Dielectric properties of the membrane of excised erythrocytes, Science, 125:985.

Schwan, H., and Kay, C. (1956). Specific resistance of body tissues, Circulation Res., 4:664.

Schwan, H., and Li, K. (1953). Capacity and conductivity of body tissues at ultrahigh frequencies, Proc. IRE, 41:1735.

Schwan, H., and Li, K. (1955a). Measurement of materials with high dielectric constant and conductivity at ultrahigh frequencies, Trans. AIEE, Communication and Electronics, 74:603.

Schwan H., and Li, K. (1955b). Measurement of materials at ultrahigh frequencies, Elec. Eng., 74:64.

Schwan, H., and Li, K. (1956a). Hazards due to total body irradiation by radar, Proc. IRE, 44:1572.

Schwan, H., and Li, K. (1956b). Mechanism of absorption of ultrahigh-frequency electromagnetic energy in tissues as related to the problem of tolerance dosage, Trans. IRE, Med. Electronics, PGME-4:45.

Schwan, H., and Piersol, G. (1954). Special review: The absorption of electromagnetic energy in body tissues. Part I. Biophysical aspects, Am. J. Phys. Med., 33:371.

Schwan, H., and Piersol, G. (1955). The absorption of electromagnetic energy in body tissues. A review and cirtical analysis. Part II. Physiological and clinical aspects, Am. J. Phys. Med., 34:425.

Schwan, H., and Sittel, K. (1953a). Wheatstone bridge for admittance determinations of highly conducting materials at low frequencies, Trans. AIEE, Communication and Electronics, 72:114.

Schwan, H., and Sittel, K. (1953b). Wheatstone bridge for admittance determinations, Elec. Eng., 72:483.

Schwan, H., et al. (1954). Electrical resistivity of living body tissues at low frequencies, Federation Proc., 13:131.

Searle, G., et al. (1961). Effects of 2450 Mc microwaves in dogs, rats, and larvae of the common fruit fly, in: Biological Effects of Microwave Radiation, Vol. 1, Plenum Press, New York, p. 187.

Seguin, L. (1949a). Lois de la répartition de la chaleur dans les tissus de l'organisme après irradiation par un champ de microondes Compt. rend., 228:135.

Seguin, L. (1949b). Réversibilité des lésions observés sur de petits animaux exposés a des ondes d'ultrahaute fréquence (longueur d'onde 21 cm), Compt. Rend., 225:76.

Seguin, L., and Castelain, G. (1947a). Action des ondes d'ultrahaute fréquence sur la temperature de petits animaux de laboratoire, Compt. Rend., 224:1662.

Seguin, L., and Castelain, G. (1947b). Lésions anatomiques observés sur les animaux de laboratoire exposés a des ondes d'ultrahaute fréquence (longueur d'onde 21 cm), Compt.Rend. 224:1850.

Seguin, L., et al. (1948). Augmentation de la vitesse d'extension de cultures de tissus irradiées par des microondes (longueur d'onde 21 cm), Compt. Rend., 227:783.

Seguin, L., et al. (1949). Action spécifique des microoondes sur les cultures de tissus, J. Radiol. Electrol., 30:566.

Seipell, H., and Morrow, R. (1960). The magnetic field accompanying neuronal activity of the nervous system, J. Wash. Acad. Sci., 50:1.

Setlow, R., and Pollard, E. (1962). Molecular Biophysics, Addison-Wesley Publishing Co., Reading, Massachusetts.

Shakhbazov, V. G., et al. (1966). The effect of a constant magnetic field on the manifestation of inbred depression and heterosis,

in: Proceedings of Conference on the Effect of Magnetic
Fields on Biological Objects, Moscow, p. 84.

Shapiro, D. N. (1955). Calculation of the efficiency of screening
chambers, Radiotekhnika, 10:36.

Shaw, T., and Windle, J. (1950). Microwave techniques for measurement of the dielectric constant of fibers and films of
high polymers, J. Appl. Phys., 21:956.

Shcherbak, A. E. (1936). Principal Works on Physiotherapy, Leningrad.

Shcherbinovskii, N. S. (1964). The cyclical activity of the sun and
the concomitant rhythms of mass multiplication of organisms,
in: Life in the Universe, Mysl', Moscow, p. 400.

Shklovskii, I. S. (1953). Radio Astronomy, Gos. Izd. Tekhniko-
Teor. Lit.

Shmal'gauzen, I. I. (1964). The Regulation of Morphogenesis in
Individual Development, Nauka, Moscow.

Shmelev, V. P. (1964a). The state of electric activity of the brain
due to action of electromagnetic vibrations of the audio- and
radio-frequency range on the organism, in: Some Questions
of Physiology and Biophysics, Voronezh, p. 98.

Shmelev, V. P. (1964b). The effect of an electromagnetic field of
the audio- and radio-frequency ranges on the reflex activity
of the spinal cord, in: Some Questions of Physiology and
Biophysics, Voronezh, p. 89.

Shnol', S. E. (1965). Synchronous conformational vibrations of
molecules of actin, myosin, and actomyosin in solutions, in:
Molecular Biophysics, Nauka, Moscow, p. 56.

Shnol', S. E. (1967). Conformational vibrations of molecules, in:
Vibrational Processes in Biological and Chemical Systems,
Nauka, Moscow.

Shternberg, I. B. (1966). The effect of a constant magnetic field on
the production of specific antibodies, in: Proceedings of
Conference on the Effect of Magnetic Fields on Biological
Objects, Moscow, p. 90.

Shul'ts, N. A. (1964). The effect of solar activity on the white
blood count, in: Life in the Universe, Mysl', Moscow, p. 382.

Simonov, P. V. (1966). What is Emotion?, Nauka, Moscow.

Sinisi, L. (1954). EEG after radar application, Electroencephalog.
Neurophysiol., 6:535.

Skurikhina, L. A. (1961). The therapeutic application of micro-

waves (SHF electromagnetic field), Vopr. Kurortol., Fizioterapii i Lecheb. Fiz. Kul't., No. 4:338.

Skurikhina, L. A. (1962). Clinical and physiological bases of microwave therapy, Novosti Med. Tekhn. No. 3:9.

Smirnova, M. I., and Sadchikova, M. N. (1960). The use of radio-
active iodine to determine the functional activity of the thyroid in people working with SHF generators, in: The Biological Action of Radio-Frequency Electromagnetic fields,
Moscow, p. 50.

Soal, S., and Bateman, F. (1954). Modern Experiments in Telepathy, Yale University Press, New Haven, Connecticut.

Sokolov, V. V., and Chulina, N. A. (1964). State of the peripheral
blood due to action of radiowaves of various frequencies on
the organism, in: The Biological Action of Radio-Frequency
Electromagnetic Fields, Moscow, p. 122.

Solov'ev, N. A. (1962). Differentiation of the action of an alternating magnetic field and the emfs and currents induced by it
in living organisms, in: Proceedings of the Second All-Union
Conference on the Use of Radioèlectronics in Biology and
Medicine, Moscow, p. 29.

Solov'ev, N. A. (1963a). Mechanism of the biological action of a
pulsed electromagnetic field, Dokl. Akad. Nauk SSSR, 149:438.

Solov'ev, N. A. (1963b). Responses of the entire living organism
to an electromagnetic field, Tr. Vses. Nauch.-Issled. Inst.
Med. Inst. i Oborudovaniya, 3:120.

Solov'ev, N. A. (1963c). Action of a high-voltage electric field of
50-2000 Hz on white mice and Drosophila, in: Proceedings
of Conference on Labor Hygiene and the Biological Action of
Radio-Frequency Electromagnetic Fields, Moscow, p. 91.

Stratton, J. (1941). Electromagnetic Theory, International Series
in Physics.

Subbota, A. G. (1957a). Some tissue reactions due to local exposure to a SHF field, Tr. VMOLA, 73:165.

Subbota, A. G. (1957b). The effect of a SHF electromagnetic field
on the higher nervous activity of dogs, Tr. VMOLA, 73:35.

Subbota, A. G. (1958). The effect of a pulsed SHF electromagnetic
field on the higher nervous activity of dogs, Byul. Eksperim.
Biol. i Med., No. 10:55.

Szabo, T. (1962). Spontaneous electrical activity of cutaneous receptors in mormyrids, Nature, 194:600.

Szent-Gyorgyi, A. (1957). Bioenergetics, Academic Press, New York.

Szent-Gyorgyi, A. (1960). Introduction to a Submolecular Biology, Academic Press, New York.

Tallarico, R., and Ketchum, J. (1959). Effect of microwaves on certain behavior patterns of the rat, in: Proc. Third Tri-Serv. Conf. on Biological Effects of Microwave Energy, Rome, New York.

Tamm, I. E. (1957). Fundamentals of the Theory of Electricity, Gos. Izd. Tekhniko-Teor. Lit., Moscow.

Tarchevskii, I. A. (1964). Change in photosynthetic carbon metabolism as a nonspecific response to the action of electric factors, in: Proceedings of Concluding Scientific Conference of Kazan State University, Kazan, p. 30.

Tatarinov, V. V., and Frenkel', G. L. (1939). An Introduction to the Study of UHF, Medgiz, Leningrad.

Tchijevsky (Chizevskii), A. (1940). Research on the electrical factor of atmospheric air maintaining the life of animals, Ac. Colombina de Cienc. Exact. Fisicas y Naturales (Bogota), 4:182.

Teixeira-Pinto, A., et al. (1960). The behavior of unicellular organisms in an electromagnetic field, Exptl. Cell Res.,20:548.

Thomson, P. (1910). A physiological effect of an alternating magnetic field, Proc. Roy. Soc. (London), 82:396.

Thomson, R., et al. (1965). Modification of x-irradiation lethality in mice by microwaves (radar), Radiation Res., 24:631.

Tolgskaya, M. S., et al. (1959). Morphological changes in animals due to experimental exposure to 10-cm waves, Vopr. Kurortol., Fizioterapii i Lecheb. Fiz. Kul't., No. 1:21.

Tolgskaya, M. S., et al. (1960). Morphological changes in experimental animals due to action of pulsed and continuous SHF, in: The Biological Action of Superhigh Frequencies, Moscow, p. 90.

Tolgskaya, M. S., and Gordon, Z. V. (1960). Change in receptor and interceptor apparatus due to action of SHF, in: The Biological Action of Superhigh Frequencies, Moscow, p. 99.

Tolgskaya, M. S., and Nikonova, K. V. (1964). Histological changes in organs of white rats due to chronic exposure to high-frequency electromagnetic fields, in: The Biological Action of Radio-Frequency Electromagnetic Fields, Moscow, p. 89.

Tolgskaya, M. S., and Gordon, Z. V. (1964). Comparative morphological characterization of action of microwaves of various ranges, in: The Biological Action of Radio-Frequency Electromagnetic Fields, Moscow, p. 80.

Tomberg, V. (1961). Specific thermal effects of high-frequency fields, in: Biological Effects of Microwave Radiation, Vol. 1, Plenum Press, New York, p. 221.

Toroptsev, I. V., and Garganeev, G. P. (1965). A morphological characterization of the changes in experimental animals due to continuous prolonged action of a constant magnetic field, in: Questions of Hematology, Radiobiology, and the Biological Action of Magnetic Fields, Tomsk, p. 345.

Toroptsev, I. V., et al. (1966a). Pathomorphological changes in experimental animals due to constant and variable magnetic fields, Proceedings of Conference on Effect of Magnetic Fields on Biological Objects, Moscow, p. 72.

Toroptsev, I. V., et al. (1966b). The action of constant magnetic fields on the embryonic and postembryonic development of frogs, in: Proceedings of Conference on the Effect of Magnetic Fields on Biological Objects, Moscow, p. 73.

Trabka, J. (1963). High-frequency components in brain wave activity, Electroencepholog. Clin. Neurophysiol., 14:453.

Tul'skii, S. V., et al. (1965). The piezoelectric resonance spectra of biopolymers, in: Molecular Biophysics, Nauka, Moscow, p. 41.

Turlygin, S. Ya. (1937). The action of centimeter waves on the central nervous system, Dokl. Akad. Nauk SSSR, 17:19.

Turlygin, S. Ya. (1942). Irradiation of the human organism with 2-mm microwaves, Byul. Eksperim. Biol. i Med., No. 4:63.

Turlygin, S. Ya. (1952). Introduction to General Radio-Engineering, Gosenergoizdat, Moscow.

Tverskoi, P. N. (1962). Course in Meteorology, Gidrometizdat, Moscow.

Tyagin, N. V. (1957). Change in the blood of animals due to a SHF field, Tr. VMOLA, 73:116.

Tyagin, N. V. (1958). The thermal action of a SHF electromagnetic field, Byul. Eksperim. Biol. i Med., No. 8:67.

Umanskii, D. M. (1965). Effect of a magnetic field on the dielectric constant of technical water, Zh. Tekhn. Fiz., 35:2245.

Ummersen, C. Van (1961). The effect of 2450 Mc radiation on the development of the chick embryo, in: Biological Effects of Microwave Radiation, Vol. 1, Plenum Press, New York, p. 201.

Ummersen, C. Van, and Cogan, F. (1965). Experimental microwave cataracts, Arch. Environ. Health, 11:177.

Valentinuzzi, M. (1962). Theory of magnetophosphenes, Am. J. Med. Electronics, 1:112.

Valentinuzzi, M. (1964). Rotational diffusion in a magnetic field and its possible magnetobiological implications, in: Biological Effects of Magnetic Fields, Vol. 1, Plenum Press, New York, p. 63.

Valfre, F., et al. (1964). La sensibilita de organismi animali alle variabili cosmiche. Prove effectuate con acqua normale e con acqua fisicamente attivata, Geofis. Meteorol., 13:76.

Valitov, N. V., and Sretenskii, N. V. (1958). Radio Measurements at Superhigh Frequencies, Voenizdat, Moscow.

Vasil'ev, L. L. (1962a). Suggestion at a Distance, Gospolitizdat, Moscow.

Vasil'ev, L. L. (1962b). Experimental Investigations of Suggestion, Izd. Leningr. Gos. Univ., Leningrad.

Vasil'ev, N. V. (1965). Effect of constant and alternating magnetic fields on the immunobiological response of the organism, in: Questions of Hematology, Radiobiology, and the Biological Action of Magnetic Fields, Tomsk, p. 379.

Vendrik, A., and Vos, J. (1958). Comparison of the stimulation of the warmth sense organ by microwave and infrared, J. Appl. Physiol., 13:435.

Vernadskii, V. I. (1926). The Biosphere. First and Second Essays, Nauchno-Tekhn. Izd., Leningrad.

Vibrational Processes in Biological and Chemical Systems (1967). Nauka, Moscow.

Villee, C. (1952). Biology, Saunders.

Vogelhut, P. (1960). Study of enzymatic activity under the influence of 3-cm electromagnetic radiation, in: Third Internat. Conf. Med. Electronics, p. 52.

Vogelman, J. (1961). Microwave instrumentation for the measurement of biological effects, in: Biological Effects of Microwave Radiation, Vol. 1, Plenum Press, New York, p. 23.

Vol'kenshtein, M. V. (1965). Molecules and Life, Nauka, Moscow.

Volkers, W., and Candib, B. (1960). Detection analysis of high-frequency signals from muscular tissues with ultra-low-noise amplifiers, IRE Int. Convent. Rec., Part 9, 8:116.

Wasserman, G. (1956). An outline of a field theory of organismic form and behaviour, in: Ciba Foundation Symposium on Extrasensory Perception, J. Churchill, London.

Watanabe, A., and Bullock, P. (1960). Modulation of activity of one neuron by subthreshold slow potentials in another in lobster cardiac ganglion, J. Gen. Physiol., 43:1031.

Webb, H., et al. (1959). Effects of imposed electrostatic field on rate of locomotion in Ilyanassa, Biol. Bull., 117:430.

Webb, H., et al. (1961). Organismic response to differences in weak horizontal electrostatic fields, Biol. Bull., 121:413.

Weiss, P. (1959a). Cellular dynamics, Rev. Mod. Phys., 31:11.

Weiss, P. (1959b). Interaction between cells, Rev. Mod. Phys., 31:449.

Wever, R. (1967). Über die Beeinflussung der circadianen Periodik des Menschen durch schwache elektromagnetische Felder, Z. Vergleich. Physiol., 56:111.

Wever, R. (1968). Einfluss schwacher elektromagnetisher Felder, Z. auf die circadiane Periodik des Menschen, Naturwissenschaften, 55:29.

Wiener, N. (1950). The Human use of Human Beings, Cybernetics and Society, Eyre and Spottiswoode, London.

Wiener, N. (1963). New Chapters in Cybernetics [Russian translation], Il, Moscow.

Wieske, C. (1963). Human sensitivity to electric fields, in: Biomedical Sciences Instrumentation, Vol. 1, Plenum Press, New York, p. 467.

Wildervank, A., et al. (1959). Certain experimental observations on a pulsed diathermy machine, Arch. Phys. Med., 40:45.

Wiley, R., et al. (1964). Magnetic reactivation of partially inhibited trypsin, in: Biological Effects of Magnetic Fields, Vol. 1, Plenum Press, New York, p. 255.

Williams, D., et al. (1956). Biological effects studies on microwave radiation. Time and power thresholds for the production of lens opacities by 12.3 cm microwaves, Trans. IRE, Med. Electronics, PGME-4:17.

Wiltschko, W. (1968). Über den Einfluss statischer Magnetfelder auf die Zugorientierung der Rotkehlchen (Erithacus rubecula), Z. Tierpsych., 25:537.

Wiltschko, W., and Merkel, F. W. (1966). Orientierung zugunruhiger Rotkehlchen im statischen Magnetfeld, Zool. Ang. Suppl.,

29:362.

Windle, J., and Shaw, T. (1954). Dielectric properties of wool-water system at 3000 and 9300 Mc, J. Chem. Phys., 22:1752.

Windle, J., and Shaw, T. (1956). Dielectric properties of wool-water system at 26,000 Mc, J. Chem. Phys., 25:435.

Wooldridge, D. (1963). Machinery of the Brain, McGraw-Hill, New York.

Yanovskii, B. M. (1964). Terrestrial Magnetism, I, Izd. Leningr. Gos. Univ.

Yeagley, H. (1947). A preliminary study of a physical basis of bird navigation. Part I, J. Appl. Phys., 18:1035.

Yeagley, H. (1951). A preliminary study of a physical basis of bird navigation. Part II, J. Appl. Phys., 22:746.

Zabotin, A. I. (1965). The effect of magnetic and electric fields on the rate and chemistry of photosynthesis, in: Questions of Hematology, Radiobiology, and the Biological Action of Magnetic Fields, Tomsk, p. 323.

Zabotin, A. I., and Neustroeva, S. I. (1966). The effect of a magnetic field on photosynthesis, in: Proceedings of Conference on the Effect of Magnetic Fields on Biological Objects, Moscow, p. 31.

Zenina, I. N. (1964). The effect of pulsed SHF electromagnetic fields on the central nervous system in the case of single and chronic exposures, in: The Biological Action of Radio-Frequency Electromagnetic Fields, Moscow, p. 26.

Zubkova, S. M. (1967a). Author's Abstract of Candidate's Dissertation: Reaction of Excitable System of Paramecia to Microwave Irradiation, Moscow.

Zubkova, S. M. (1967b). Effect of electromagnetic fields on the regulation of the motor functions of paramecia, Tr. Mosk. Obshchest. Isp'tatel. Prirod. (in press).

Zyryanov, P. S. (1961). The nature of the forces of interaction between chromosomes, Biofizika, 6:495.

Index